DISCARDED

```
536.23                                    c.1
B516

Berman, Robert
     Thermal conduction in solids
```

```
536.23                                    c.1
B516

Berman, Robert
     Thermal conduction in solids
```

COLLEGE OF THE SEQUOIAS
LIBRARY

OXFORD STUDIES IN PHYSICS

GENERAL EDITORS

B. BLEANEY, D. W. SCIAMA, D. H. WILKINSON

THERMAL CONDUCTION IN SOLIDS

BY
R. BERMAN

CLARENDON PRESS · OXFORD
1976

Oxford University Press, Walton Street, Oxford OX2 6DP

OXFORD LONDON GLASGOW NEW YORK
TORONTO MELBOURNE WELLINGTON CAPE TOWN
IBADAN NAIROBI DAR ES SALAAM LUSAKA ADDIS ABABA
KUALA LUMPUR SINGAPORE JAKARTA HONG KONG TOKYO
DELHI BOMBAY CALCUTTA MADRAS KARACHI

ISBN 0 19 851429 8

© Oxford University Press 1976

All rights reserved. No part of this publication may be reproduced, stored in a retrieval system, or transmitted, in any form or by any means electronic, mechanical, photocopying, recording or otherwise, without the prior permission of Oxford University Press

Filmset in Northern Ireland
at the Universities Press, Belfast
Printed in Great Britain
at the University Press, Oxford
by Vivian Ridler
Printer to the University

TO MAUREEN

PREFACE

My aim in writing this book has been to describe the range of phenomena involved in thermal conduction, without using any sophisticated theoretical techniques. I have had in mind the needs of three groups of readers: undergraduates who want something more than the usual brief treatments of thermal conductivity in text-books on solid state physics; research workers who intend to enter this or a related field and want an idea of the scope of the subject; and people working in quite different branches of science who encounter problems involving thermal conductivity.

My own research in this subject was initiated about 27 years ago by the late Professor Sir Francis Simon's suggestion that measurements of thermal conductivity could be used to investigate lattice defects in non-metals, in the same way that electrical conductivity was being used to study defects in metals. Over the whole of the period since then I have been extremely fortunate in the colleagues with whom I have worked. They are too numerous to mention here, but most of them are referred to in the text. I thank them all and hope that I have not misrepresented the results of their collective wisdom.

I am grateful to all authors and publishers who have given me permission to reproduce illustrations.

I am especially indebted to Dr. John Rogers who read the manuscript and made many very valuable comments and suggestions for improvements, as a result of which much that was unclear has been changed. I am, however, responsible for all the faults which remain.

Oxford R.B.
October 1975

ACKNOWLEDGEMENTS

FIGS 2.2, 7.1, 7.13, 9.3, 12.7 and 12.8 are by kind permission of the Institute of Physics, © Institute of Physics 1952, 1957, 1973, 1975. I thank the following publishers for permission to reproduce material: Academic Press for Fig. 12.1; Plenum Publishing Corporation, New York, for Fig. 8.3; the National Research Council of Canada for Fig. 8.7a; the Royal Society for Figs 7.12, 8.1, 8.2, 8.7d and 13.3; Pergamon Press for Fig. 7.6; Taylor and Francis for Figs 7.10, 9.4, 10.2, 10.4 and 12.5; North-Holland Publishing Company for Figs 8.7b and 8.7c; the American Physical Society for Figs 7.8, 7.9, 7.14, 8.4, 8.5, 8.8, 8.9, 9.1, 9.2, 11.4, 12.3, 12.4 and 13.1; the Academy of Sciences, USSR for Fig. 12.6.

CONTENTS

1. INTRODUCTION 1

2. DEFINITIONS AND PRINCIPLES OF METHODS OF MEASUREMENT 3
   2.1. Definitions 3
   2.2. Measurement of thermal conductivity 3
      2.2.1. Steady-state methods 5
      2.2.2. Non-steady-state methods 7
      2.2.3. The temperature range of thermal-conductivity measurements .. 9

3. GENERAL BEHAVIOUR OF THE THERMAL CONDUCTIVITY OF METALS AND OF NON-METALLIC CRYSTALS 11
   3.1. Metals 11
   3.2. Non-metallic crystals 13
   3.3. Comparison between metals and non-metals 14

4. PHONONS AND THE BOLTZMANN EQUATION 16
   4.1. Phonons 16
      4.1.1. Vibrations of a discrete lattice 16
      4.1.2. Quantization of vibrational energy 19
   4.2. The Boltzmann equation 20
      4.2.1. The relaxation-time method 21
      4.2.2. The variational method 23

5. NORMAL AND UMKLAPP PROCESSES 29
   5.1. Phonon–phonon interactions 29
      5.1.1. Normal and Umklapp processes 30
   5.2. The effect of N-processes 32

6. TAKING ACCOUNT OF NORMAL PROCESSES 36
   6.1. The relaxation-time method 36
      6.1.1. Resistive processes dominant 37
      6.1.2. Resistive processes present, but N-processes dominant 38
      6.1.3. Only N-processes acting 39
   6.2. The variational method 40
      6.2.1. Resistive processes and N-processes both important 40
      6.2.2. Resistive processes present, but N-processes dominant 42
   6.3. The method of Guyer and Krumhansl 43

7. THE THERMAL CONDUCTIVITY OF NEARLY PERFECT NON-METALLIC CRYSTALS 45
   7.1. U-processes 45
      7.1.1. High temperatures 46
      7.1.2. Low temperatures 56
   7.2. Boundary scattering 61
   7.3. Poiseuille flow of phonons 69

CONTENTS

8. THE THERMAL CONDUCTIVITY OF IMPERFECT CRYSTALS . 73
- 8.1. Phonon scattering by defects ... 73
  - 8.1.1. Point defects ... 73
  - 8.1.2. Larger defects ... 76
  - 8.1.3. Dislocations ... 77
- 8.2. Combining intrinsic and defect scattering rates ... 79
  - 8.2.1. Dominant phonon method ... 80
  - 8.2.2. The 'Debye approximation' ... 81
  - 8.2.3. The variational method ... 82
- 8.3. Thermal conductivity of crystals containing defects ... 82
  - 8.3.1. Non-resonant point defects ... 83
  - 8.3.2. Resonant scattering ... 91
  - 8.3.3. Larger defects ... 99
  - 8.3.4. Dislocations ... 101

9. AMORPHOUS SOLIDS ... 104
- 9.1. General behaviour of the conductivity ... 104
- 9.2. The 'high'-temperature behaviour ... 109
- 9.3. The conductivity at very low temperatures ... 110

10. ELECTRONS ... 115
- 10.1. Drude theory of conductivity ... 116
- 10.2. Electrons in metals ... 117
  - 10.2.1. Free electrons ... 117
  - 10.2.2. Electrons in a crystal lattice ... 120
  - 10.2.3. Metals, insulators, and semiconductors ... 121
- 10.3. Conduction by electrons ... 123
- 10.4. Conditions for the validity of the Wiedemann–Franz–Lorenz law ... 125

11. ELECTRON SCATTERING ... 128
- 11.1. Phonon–electron scattering ... 128
- 11.2. The temperature variations of the resistivities ... 134
  - 11.2.1. The 'ideal' electrical resistivity ... 134
  - 11.2.2. The 'ideal' electronic thermal resistivity ... 134
  - 11.2.3. The phonon thermal resistivity due to electron scattering ... 135
- 11.3. Some corrections to the simple theory ... 136
  - 11.3.1. Electron–phonon U-processes ... 136
  - 11.3.2. Electron–electron scattering ... 137
  - 11.3.3. Electron–phonon scattering when the electron mean free path is small ... 138
- 11.4. Scattering by defects ... 140

12. THERMAL CONDUCTIVITY OF METALS AND ALLOYS ... 143
- 12.1. Pure metals ... 143
  - 12.1.1. Estimates of the lattice component ... 144
  - 12.1.2. The electronic component ... 145
- 12.2. Alloys ... 151
  - 12.2.1. The electronic contribution ... 152
  - 12.2.2. The lattice conductivity in metals and alloys ... 154
  - 12.2.3. The dependence of lattice conductivity on electron mean free path ... 158
  - 12.2.4. The influence of dislocations on the lattice conductivity of alloys ... 161
- 12.3. Superconductors ... 164

13.	SEMICONDUCTORS	169
	13.1. Pure intrinsic semiconductors	169
	13.2. Impure semiconductors	174
	13.2.1. Semiconductor alloys	174
	13.2.2. Doped semiconductors	175
	REFERENCES	179
	AUTHOR INDEX	187
	SUBJECT INDEX	191

1
INTRODUCTION

THE applications of an understanding of thermal conduction have increased with the growth of technologies involving operations at both high and low temperatures. The processes concerned in conduction have also become better understood during the last 25 years and can serve as a guide for choosing materials with the desired conduction properties, although accurate quantitative predictions can seldom be made.

There are several mechanisms by which heat can be transmitted through a solid and many processes which limit the effectiveness of each mechanism. In a non-metal heat is conducted by means of the thermal vibrations of the atoms. In a simple metal this mode of heat transport makes some contribution, but the observed thermal conductivity is almost entirely due to the electrons. It does not follow that the thermal conductivities of the two types of solid must be very different in magnitude, as they are in the case of electrical conductivity, but their dependences on temperature and on the imperfections in individual specimens are rather different. In many solids, such as alloys and semiconductors, both transport mechanisms can make comparable contributions to the observed conductivity, and the relative proportions vary with composition and temperature. In superconductors the proportions are different in the normal and superconducting states, so that below the transition temperature they can be changed by an appropriate magnetic field.

The aim of this book is to explain the main mechanisms by which heat is conducted in solids and to give relatively simple examples in illustration. Many of the experimental results quoted are of measurements made below room temperature. This is a deliberate choice because the interpretation of such experiments is generally more sensitive to the details of the theory. There is no attempt to delve deeply into the theory, since any worthwhile treatment would at least double the length of the book. Theoretical developments are still taking place, but the sophistication reached nearly 20 years ago can be judged from, for example, the review article by Klemens (1958) or the book by Ziman (1960). More recent advances in phonon theory are reviewed by Beck, Meier, and Thellung (1974). Similarly, detailed descriptions of the interpretations of particular experimental results are, on the whole, omitted to avoid an indigestible mass of detail which would obscure the more generally applicable principles. Such details can be found in many of the papers to which reference is made.

The treatments of non-metals and of metals follow parallel paths. The outline of experimental methods (Chapter 2) is relevant to both, and so is the brief resumé of the gross features of the conductivities (Chapter 3). Phonons, which are the carriers of heat in non-metals, are then discussed, followed by a consideration of the ways in which phonons are scattered and of the conductivity to which they give rise (Chapters 4 to 8). Non-crystalline solids, such as glasses, are treated separately (Chapter 9). Chapters 10 to 12 are concerned with the properties of electrons in metals, their scattering, and the thermal conductivities contributed by phonons and electrons in metals and alloys. Because much of the theory of electrons is more familiar from its bearing on the electrical conductivity of metals, the discussion here is shorter than for phonons. Only brief mention is made of the thermal conductivity of superconductors. Semiconductors, in which both the lattice and the electronic thermal conductivities are important, are treated in Chapter 13.

2
DEFINITIONS AND PRINCIPLES OF METHODS OF MEASUREMENT

2.1. Definitions

IN AN isotropic solid heat flow obeys the relation

$$\mathbf{h} = -\kappa \operatorname{grad} T \qquad (2.1)$$

where $\mathbf{h}$ is a vector measuring the rate of flow of heat across unit cross-section perpendicular to $\mathbf{h}$, T is the temperature, κ is the thermal conductivity, and the negative sign indicates that heat flows down a temperature gradient from the hotter to the colder regions.

In a crystal which does not have cubic symmetry, $\mathbf{h}$ is not in general parallel to grad T and eqn (2.1) is replaced by

$$h_i = -\kappa_{ij} \frac{\partial T}{\partial x_j} \qquad (2.2)$$

where the coefficients κ_{ij} form a second-rank tensor. When referred to the principal axes, eqn (2.2) has the same form as eqn (2.1), and the conductivities in these directions are usually quoted for anisotropic crystals.

At normal temperatures, κ is independent of the shape and size of the specimen measured, so that a unique value can be defined at any particular temperature. However, for every non-metallic crystal of normal purity and perfection there is some temperature below which the conductivity, as deduced from normal measurements, is size dependent (see § 7.2). For crystals with their smallest dimensions of the order of millimetres this temperature ranges from ~ 1 K for low-density helium to ~ 100 K for diamond.

The S.I. unit of thermal conductivity is Watt-metre^{-1}-Kelvin^{-1} (W m^{-1} K^{-1}), and values expressed in these units are 100 times greater than those given with the centimetre as the unit of length (W cm^{-1} K^{-1});

$$1 \text{ S.I. unit} = 5 \cdot 78 \times 10^{-1} \text{ B.T.U. ft}^{-1} \text{ h}^{-1} \deg \text{F}^{-1}.$$

2.2. Measurement of thermal conductivity

In the simplest steady-state experimental arrangement, illustrated schematically in Fig. 2.1, heat is supplied at one end of a rod of uniform

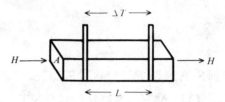

FIG. 2.1. Principle of the steady-state longitudinal heat-flow method for measuring thermal conductivity.

cross-sectional area A at a known rate H and is removed at the other end. Thermometers are attached at two places along the specimen separated by a distance L, and the temperature difference ΔT between them is measured. The conductivity is then derived from the relation

$$H = \kappa A \frac{\Delta T}{L}. \qquad (2.3)$$

If ΔT is not too large, the derived value of κ will be that corresponding to the mean temperature between the thermometers even if κ is a rapidly varying function of temperature. As an example, assume that the conductivity is given by $\kappa = KT^3$ and that the temperature difference along unit length of the specimen of unit cross-sectional area is ΔT, the lower temperature being T. Then the heat flow is

$$\int_T^{T+\Delta T} \kappa \, dT = K \int_T^{T+\Delta T} T^3 \, dT.$$

The thermal conductivity is derived from the measurement by dividing the heat flow by the temperature difference and is thus $\frac{1}{4}K\{(T+\Delta T)^4 - T^4\}/\Delta T$. The mean temperature between the thermometers is $T+\Delta T/2$, and at this temperature the true conductivity is $K(T+\Delta T/2)^3$. The difference between these expressions is $\Delta \kappa \approx T(\Delta T)^2/4$ and

$$\Delta \kappa / \kappa \approx \tfrac{1}{4}(\Delta T/T)^2.$$

Even for a relative temperature difference as large as $\Delta T/T = \frac{1}{10}$, the difference between the apparent and true conductivities is only $\frac{1}{4}$ per cent.

The steady-state longitudinal heat-flow method can be used in just the form illustrated if it is certain that nearly all the heat supplied by the heater actually travels through the specimen to the colder end. Although heat exchange through the surrounding medium and along electrical leads can be made small by careful design and by making measurements in vacuo, heat exchange by radiation effectively limits the temperature up to which this simple method can be used. For a very poor conductor, such as glass, the radiation heat loss may be too large to be accounted for accurately even below liquid-air temperature, while for a good conductor it may be possible to work up to room temperature without serious trouble.

The upper temperature limit can be extended by the use of radiation shields along which there is the same temperature distribution as exists along the specimen. However, the steady-state method can best be extended by heating the specimen internally so that the heat cannot be lost by radiation until it has all passed through the specimen and has brought about the corresponding temperature gradient.

Many determinations of thermal conductivity at high temperatures are now made by non-steady-state methods. Either the rate of change of an unsteady temperature distribution or the amplitude and phase relations in a temperature wave are determined. From such measurements the ratio of conductivity to specific heat can be derived.

2.2.1. Steady-state methods

2.2.1(a). Longitudinal flow. In many early measurements the specimen was held between a heater and a heat sink, and the temperature difference between these was measured. However, there can be a considerable thermal resistance at the junction between two different materials, so that there would be temperature discontinuities at the ends of the specimen. For a good conductor the contact resistance can be hundreds of times greater than the thermal resistance of the specimen itself, so that the overall resistance measured bears little relation to the desired specimen resistance. For accurate results it is therefore necessary to make the thermal measurement analogous to the potentiometric method for determining electrical resistance. Measurement of thermal conductivity is in general more difficult, however, because it is more difficult to arrange for negligible heat flow away from the specimen along the heater and thermometer leads than it is to achieve negligible current flow along electrical leads to a potentiometer.

Problems which arise in using this method at low temperatures have been discussed by Berman (1961) and White (1969), while Kopp and Slack (1971) considered in detail the precautions necessary when using thermocouples to measure the temperature difference along the specimen. High-temperature measurements are described by Laubitz (1969).

2.2.1(b). Radial heat flow. If heat is supplied internally to a specimen, then radiation and other heat losses certainly determine the steady-state temperature at the surface, and thus the temperature throughout the specimen, but the measured temperature difference has been caused by the total measured heat input (apart from losses from leads if conventional heaters are used).

If a heater is inserted along the axis of a hollow cylinder and provides heat uniformly along its length, the heat flow per unit length from the inside to the outside is related to the temperatures T_{r_1} and T_{r_2} measured at radii r_1 and r_2 from the axis by the equation

$$H = \frac{2\pi\kappa(T_{r_1} - T_{r_2})}{\ln(r_2/r_1)} \text{ per unit length}$$

for the parts of the rod sufficiently far from the ends for longitudinal heat flow to be negligible. Longitudinal flow can be reduced by making the specimen in the form of a stack of discs; this should not change the radial flow, but the longitudinal flow is impeded by the thermal resistance at each interface.

Internal heating is also used with spherical and ellipsoidal specimens, and for a non-metallic specimen the heater can be a metal in which eddy-current heating is produced by an external high-frequency coil. A modest advantage of using a cylinder is that only the ratio of two radii is required, and in calculating the absolute value of the thermal conductivity a knowledge of the thermal expansion of the specimen is not needed. A general discussion of radial methods is given by McElroy and Moore (1969).

2.2.1(c). *Electrical heating of the specimen.* Heat can be generated in an electrical conductor by passing current through the specimen itself. This heat is generated inside the material, and for purely radial heat flow radiation only determines the absolute temperature at which measurements are made for a given power dissipation. For a solid cylinder of radius r_2, the temperature difference between the central axis and the surface for a portion far from the ends is given by

$$T_0 - T_{r_2} = \frac{V_\varepsilon^2 r_2^2}{4\kappa\rho}$$

If the cylinder is hollow, the central and outer temperatures can be determined with a pyrometer at high temperatures. If the inner radius is r_1, then

$$T_{r_1} - T_{r_2} = \frac{V_\varepsilon^2}{4\kappa\rho}\{r_2^2 - r_1^2(1 - 2\ln(r_2/r_1))\}$$

where in both cases V_ε is the potential drop along the conductor per unit length and both κ and ρ, the electrical resistivity, are assumed to be constant over the temperature range involved.

For purely longitudinal heat flow, as in a thin wire carrying a current, the central portion of the conductor is hotter than the ends. If the potential drop between the ends, maintained at T_end, is V'_ε and T_mid is the temperature at the centre, then, assuming no loss by radiation,

$$T_{mid} - T_{end} = \frac{(V'_e)^2}{8\kappa\rho}.$$

Details of direct heating methods are given by Flynn (1969).

2.2.1(d). Comparative methods. Both the longitudinal and radial methods lend themselves to comparative measurements. In the longitudinal heat-flow method the temperature gradients in the specimen and in a standard material through which the same heat flows are compared. If a standard is placed on either side of the specimen, then any heat loss from the sides of the specimen will be shown up by the difference between the heat flows in the two standards. In a radial heat flow arrangement, the standard would be concentric with the specimen.

2.2.2. Non-steady-state methods

In general, an arbitrary temperature distribution set up in a body will change with time at a rate dependent on the ratio of the thermal conductivity to the heat capacity per unit volume, C. This ratio is the thermal diffusivity $\mathcal{D} = \kappa/C$. In one type of application of this principle, a periodic temperature variation is imposed at one end of a rod and certain characteristics of the propagation of the resulting temperature wave are determined. Another common application is to heat one part of a body for a limited time and follow the subsequent temperature changes at other parts.

2.2.2(a). The temperature-wave method. In Ångström's method (1861) the temperature at one end of a long rod is made to vary sinusoidally (this would be the simplest type of variation) and the attenuation of the temperature excursions and their time lag along the rod are determined. If at $x = 0$ the temperature difference T' between the rod and its mean temperature varies with time as

$$T'(0) = T'_0 \sin(\omega t)$$

then for no heat losses from the sides of the rod, the temperature difference from the mean at a distance x along the rod is

$$T'(x) = T'_0 \exp\{-\sqrt{(\omega/2\mathcal{D})}x\} \sin(\omega t - \sqrt{(\omega/2\mathcal{D})}x).$$

Under the ideal conditions envisaged, the thermal diffusivity could be determined from *either* the attenuation of the wave $1/\Lambda(x) = \exp(-\sqrt{(\omega/2\mathcal{D})}x)$, *or* from the phase lag $\psi(x) = \sqrt{(\omega/2\mathcal{D})}x$, so that

$$\mathcal{D} = \frac{\omega x^2}{2(\ln \Lambda)^2} = (\omega x^2/2\psi^2).$$

It is difficult to prevent heat losses from the side of the rod except at low temperatures. If, however, the loss is proportional to the excess temperature T', as it would be if T' is not too large, then it can be shown that $\mathscr{D}$ can still be deduced, but two separate measurements must now be made, such as the velocity of propagation or the rate of amplitude decrement for two different values of ω, or both of these must be measured at a particular frequency. For example, although neither Λ nor ψ alone will give the correct value for $\mathscr{D}$ when there are heat losses, the relation

$$\mathscr{D} = \frac{(\omega x^2)}{(2\psi \ln \Lambda)}$$

still holds. Howling, Mendoza, and Zimmerman (1955) and Zavaritskii (1957) used the temperature-wave method at very low temperatures and only measured ψ, but at normal and high temperatures various authors have measured the different pairs of quantities necessary to eliminate the effect of heat loss.

Two difficulties which arise in employing the method are the establishment of a purely sinusoidal temperature variation and the gradual upward drift in the mean temperature of the rod. A neat solution of the latter problem is described by Green and Cowles (1960), who produced the heating and cooling by passing current through a junction between p-type and n-type bismuth telluride in periodically reversed directions. Because of the Peltier effect, heat was given out at the junction for one direction of current flow and absorbed for the other. The irreversible Joule heating was annulled over a cycle by passing a larger current in the direction which produced Peltier cooling. This method of heating also helps to produce a sinusoidal variation of temperature. The end of the rod in contact with the 'heater' has a temperature varying fairly symmetrically with time, which can be represented by a Fourier series which has small even harmonics. The main terms therefore have frequencies ω, 3ω, etc., but as the attenuation of waves is greater for higher frequencies, the wave becomes almost purely sinusoidal at a small distance from the heater. Its subsequent attenuation and velocity can then be determined and the values used directly to calculate $\mathscr{D}$.

2.2.2(b). *Transitory heating methods.* These have been applied to specimens of various shapes, generally arranged in such a way that the heat flow is either purely longitudinal or purely radial so that the diffusivity can be derived without too complex computations. At high temperatures radiation loss can be kept small by reducing the duration of the experiment. This is achieved by flash heating of one face of a thin plate either with a flash tube or a laser. The plate starts at a steady temperature and after the flash the temperature of the face remote from the heat source is monitored. If there is no heat loss during the time it takes for this face to reach its maximum temperature, which must be much longer than the duration of the heating pulse, there is a simple relation between the time to attain half the maximum temperature rise, $t_{\frac{1}{2}}$, the thickness of the plate, d, and the thermal diffusivity

$$\mathcal{D} = 1 \cdot 37 d^2 / (\pi^2 t_{\frac{1}{2}}).$$

Mention should be made in this section of a very rapid method for estimating conductivities near normal temperatures which was first described by Powell (1957). This thermal comparator, sketched in Fig. 2.2, consists essentially of two phosphor bronze (or other metallic) balls, held in balsa wood and heated to some temperature above (or cooled below) that of the specimen to be studied. The device is then placed on the specimen so that it rests on two studs and on one of the balls, while the other ball is free. The temperature difference between the two balls is followed by a thermocouple connected between them. It is found that under certain conditions there is a linear relation between the square root of the thermal conductivity and the rate at which the temperature difference changes.

The temperature wave and transitory methods are described by Danielson and Sidles (1969).

2.2.3. *The temperature range of thermal-conductivity measurements*

Thermal conductivities have been measured from temperatures well below 1 K to well above 3000 K. At low temperatures, measurements are usually relatively straightforward and there is good agreement (better than 10 per cent) between results found by different authors using various types of apparatus to measure the same or similar specimens. However, as the temperature increases, differences between measured values grow and become much greater than the individual experimenters' own estimates of their accuracy. This is found to occur even when comparing results obtained by similar measuring techniques. For example, McElroy

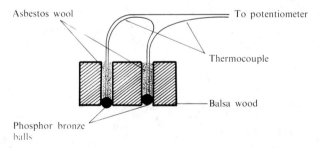

FIG. 2.2. Diagrammatic section through thermal comparator. (From Powell 1957.)

and Moore (1969) show comparisons for several materials measured by radial heat-flow methods up to about 2000 K, while Flynn (1969) gives a comparison of values for the thermal conductivity of tungsten determined by direct electrical heating methods up to 3500 K. In some cases the spread of results covers a factor of two.

An international project is being sponsored by CODATA (Committee on Data for Science and Technology) in order to resolve such discrepancies and to establish certain standard reference materials for thermal conductivity.

3
GENERAL BEHAVIOUR OF THE THERMAL CONDUCTIVITY OF METALS AND OF NON-METALLIC CRYSTALS

BEFORE embarking on the more detailed discussions of later chapters, an outline will be given of the observed behaviour of the thermal conductivities of pure metals and of pure non-metallic crystals, and their interpretation in terms of mean free paths of the appropriate carriers.

3.1. Metals

The high thermal conductivity of metals such as copper and silver is familiar in everyday life, and is closely connected with their high electrical conductivity. In the Drude theory (1900, 1902) of electrical conductivity it is assumed that there is some average distance, the mean free path l, over which free electrons are accelerated by the electric field before they lose the extra velocity acquired and resume their typical purely thermal motion. The acceleration is brought to an end by some sort of collision with the atoms. The electrical conductivity can then be expressed as

$$\sigma = \frac{n_e e^2 l}{2 m_e v} \qquad (3.1)$$

where n_e is the number of free electrons per unit volume, e and m_e are their electric charge and mass, and v is their mean thermal velocity (a slightly more sophisticated treatment gives twice this value for σ and is used in §10.1).

If it is assumed that in a temperature gradient electrons travel just the same average distance l before transferring their excess thermal energy to the atoms by collisions, then the heat conductivity is given by

$$\kappa = \tfrac{1}{3} n_e c_e v l \qquad (3.2)$$

where c_e is the heat capacity per electron ($n_e c_e$ is the electronic heat capacity per unit volume, C_e). The Wiedemann–Franz law (1853) is obtained by comparing κ with σ as given by eqns (3.1) and (3.2):

$$\frac{\kappa}{\sigma} = \frac{2}{3} \frac{m_e v^2 c_e}{e^2}.$$

On a classical theory, in which free electrons are treated as gas-like particles, c_e is constant and v^2 is proportional to the absolute temperature

T, and we obtain

$$\frac{\kappa}{\sigma} = 3\left(\frac{k_B}{e}\right)^2 T \tag{3.3}$$

where k_B is Boltzmann's constant. From a quantum statistical mechanical treatment of electrons forming a highly degenerate system, v is effectively independent of temperature, while c_e is proportional to T, and we obtain

$$\frac{\kappa}{\sigma} = \frac{\pi^2}{3}\left(\frac{k_B}{e}\right)^2 T \tag{3.4}$$

Both these treatments yield the Wiedemann–Franz–Lorenz law (Lorenz 1881) $\kappa/\sigma T$ = constant, and although in the classical treatment c_e and v are quite incorrectly given, the numerical constants in eqns (3.3) and (3.4) are very little different (the agreement is, of course, worse if the 'better' classical theory is used).

At sufficiently high temperatures the effectiveness of the lattice in scattering electrons is proportional to the lattice vibrational energy, which is proportional to T. Then l is inversely proportional to T and this is also the temperature variation of the electrical conductivity. As the temperature is decreased, collisions are less effective in limiting the mean free path and the conductivity increases faster than the $1/T$ law would predict.

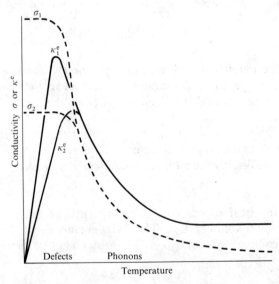

FIG. 3.1. Electrical conductivity σ and electronic thermal conductivity κ^e of a metal as functions of temperature. The dominant electron-scattering mechanisms are indicated along the abscissa. The upper curves in each case are for more perfect specimens than the lower curves.

Eventually, the mean free path reaches a constant value determined by imperfections in the lattice, such as impurities, interstitial atoms, and other defects. The electrical conductivity is then constant (the inverse of this limiting value is called the residual resistivity, ρ_0).

If we use the Wiedemann–Franz–Lorenz (WFL) law to predict the thermal conductivity of a fairly pure simple metal from the measured electrical conductivity, we obtain the correct general behaviour because almost the entire observed thermal conductivity is due to the electrons. The details are wrong because we cannot, in general, assume that the effective values of l are the same for electrical and thermal conductivities. At high temperatures the thermal conductivity is constant and at low temperatures it is proportional to the temperature, as would be deduced by adding one power of T to the temperature variation of the electrical conductivity. At intermediate temperatures, however, the thermal conductivity varies less rapidly with temperature than would be expected from the WFL law. The electrical and thermal conductivities of a pure metal are shown schematically in Fig. 3.1.

3.2. Non-metallic crystals

It is more difficult to identify the heat carriers when there are no free electrons. Because the atoms in a solid are closely coupled together, an increase in vibrational energy in one part of a crystal, associated with an increase in temperature, will be transmitted to the other parts. Debye (1914) regarded the heat as transmitted by the lattice vibrations forming a wave motion, and the appropriate mean free path as the distance such a wave travels before its intensity is attenuated by scattering to $1/e$ of its initial value. In modern theory, heat is considered as being transmitted by phonons, which are the quanta of energy in each mode of vibration, and the mean free path is a measure of the rate at which energy is exchanged between different phonon modes. We can again use the expression

$$\kappa = \tfrac{1}{3} C v l \qquad (3.5)$$

to represent the heat conductivity, where v is now the mean phonon velocity, approximately equal to the velocity of sound in the crystal, and C is the heat capacity contributed by the lattice.

At normal and high temperatures l is limited by direct interactions among the phonons themselves, and at sufficiently high temperatures l is inversely proportional to T. As the temperature decreases, interactions among the phonons become rapidly less effective in restricting l, which thus increases more rapidly than $1/T$. For sufficiently perfect crystals this increase is best represented by an exponential of the form $l \propto \exp(T^*/T)$, where T^* is a characteristic temperature for the particular crystal and is an appreciable fraction of the Debye characteristic temperature (discussed in

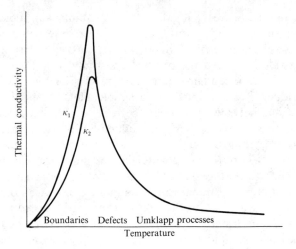

FIG. 3.2. Thermal conductivity of a non-metallic crystal. The dominant phonon-scattering mechanisms are indicated along the abscissa. The upper curve is for a crystal of larger diameter than the lower curve.

§ 4.2.1.). At easily accessible low temperatures (~ 1 to ~ 100 K, depending on the material) l may reach several millimetres and is thus comparable with the smallest dimensions of most specimens studied. It then tends to a constant value which depends on the shape and size of the specimen.

In order to deduce the behaviour of the thermal conductivity from that of the mean free path, we must introduce the temperature variation of the heat capacity, but the mean velocity can be regarded as independent of temperature. At temperatures which are high enough for the relation $l \propto 1/T$ to hold, C is nearly constant so that $\kappa \propto 1/T$. As the temperature decreases, C also decreases, eventually varying as T^3, but the exponential variation in l is then so dominant that the conductivity too is essentially represented by the exponential variation. Finally, when the mean free path becomes constant, the temperature dependence of the conductivity reflects the T^3 behaviour of the specific heat. A typical conductivity curve for a good non-metallic crystal is shown in Fig. 3.2.

3.3. Comparison between metals and non-metals

In comparing Figs 3.1 and 3.2, there are a few general remarks which can be made. At high temperatures the appropriate mean free paths are both proportional to $1/T$. For a metal the electronic heat capacity is proportional to temperature, while for a non-metallic crystal it is constant. The temperature dependences of the two thermal conductivities thus differ by one power of T.

As the maxima are approached, the mean free paths increase much more rapidly than $1/T$, and, although the rates of increase are different for metals and non-metals, in both cases this rapid change more than makes up for the decreasing heat capacity. For metals the constant thermal conductivity changes to a conductivity proportional to $1/T^2$, while for non-metals the $1/T$ variation changes to an exponential increase.

At the lowest temperatures both mean free paths reach constant values and the conductivity is proportional to the heat capacity contributed by the appropriate heat carriers. As a result, the thermal conductivity of a metal is proportional to T, while for a non-metallic crystal it is proportional to T^3. It should be emphasized, however, that for a metal the constant mean free path is determined by the imperfections present, while for a non-metal it is determined by the external boundaries of the crystal. Imperfections naturally affect lattice thermal conductivity, but this influence falls off at the lowest temperatures. As the temperature decreases, the important lattice waves are those of longer wavelength, and for an 'average' crystal (with Debye characteristic temperature between 200 and 300 K) the dominant wavelengths reach ~ 100 atomic spacings at 1 K. Such long waves are almost unaffected by disorders on an atomic scale, but are scattered on reaching the external boundaries of a crystal. On the other hand, the energies of the electrons in ordinary metals which are effective in conduction are little dependent on the temperature. Their de Broglie wavelengths in copper, for example, are at all temperatures just about equal to the interatomic spacing. Electrons are therefore strongly scattered by imperfections of atomic dimensions.

The orders of magnitude of the conductivities should also be commented on. For pure materials the values of the maximum conductivities are comparable—1000 to 20,000 W m^{-1} K^{-1} for many pure metals and non-metallic crystals—but the temperature variation on both sides of the maximum is faster for non-metals than for metals. At sufficiently high and at sufficiently low temperatures, therefore, most non-metals are distinctly poorer heat conductors than most metals. There are exceptions to this general rule: the maximum occurs at such a 'high' temperature for diamond (liquid-air temperature) that it is a far better heat conductor than copper even well above room temperature, while the maximum occurs at such a low temperature for solid helium (below 1 K for helium solidified at low pressure) that it is a far better heat conductor than copper in the region of 1 K.

4
PHONONS AND THE BOLTZMANN EQUATION

SINCE it is present to some extent in all solids, lattice thermal conduction will be considered first. In non-metals it is the only conduction mechanism or at least the dominant one over a wide temperature range, and may even dominate the measured conductivity in alloys and in superconductors.

In terms of phonons, a flow of heat implies that the phonon distribution differs from that in thermal equilibrium, which corresponds to no net heat flow. The conductivity depends on the extent to which the distribution can depart from equilibrium for a given temperature gradient. In the type of theory which will be most used in subsequent interpretations of experimental data, the departure from equilibrium is expressed in terms of relaxation times or mean free paths, which in general depend on temperature and on the frequency and polarization of the phonon mode. The relaxation times are determined by many processes, both intrinsic to the material and particular to a given specimen.

4.1. Phonons

4.1.1. Vibrations of a discrete lattice

The computation of the possible modes of vibration of a real three-dimensional solid is extremely difficult, but the essential ideas necessary for understanding the general properties of normal modes can be derived from consideration of a linear chain of atoms. It can easily be shown for a simple linear chain that there is only a limited number of really different modes of vibration which represent different coupled motions of the atoms. The variation of propagation velocity with wavelength can also be calculated exactly for this case.

The simplest linear chain consists of N identical atoms of mass M, held together by harmonic forces acting only between adjacent atoms and constrained to move along the length of the chain. The possible angular frequencies of the modes in which all atoms vibrate with the same frequency ω vary with wave-number q according to the relation

$$\omega(q) = \sqrt{\frac{\zeta}{M}} 2 \left| \sin \frac{qa}{2} \right| \qquad (4.1)$$

where ζ is the harmonic force constant and a is the distance between the

PHONONS AND THE BOLTZMANN EQUATION 17

rest positions of the atoms. This relation—the dispersion curve—is shown in Fig. 4.1, and it can be seen that it differs greatly from the dispersion curve for a continuum, which for comparison has been assumed to have the same behaviour as the discrete chain in the long-wavelength (small q) limit. In a travelling wave q determines the relative phase of the motion of neighbouring atoms, and for our chain q can take any value, either positive or negative, which is a multiple of $2\pi/Na$ (if we assume 'periodic boundary conditions', so that atom 1 is behaving in the same way as would atom $N+1$). From eqn (4.1) it can be seen that the frequencies which occur for q between 0 and π/a are repeated for negative q and also for q outside the range 0 to $\pm\pi/a$. Between $q = 0$ and $q = +\pi/a$ there are $N/2$ modes which represent a wave travelling in the positive direction and there are another $N/2$ modes for negative q. A possible mode is also represented by $q = 0$ (all atoms moving in phase, resulting in bodily motion of the chain as a whole), but, as modes $q = +\pi/a$ and $q = -\pi/a$ represent identical standing waves, we are left with exactly N values of q which represent different relative motions of the atoms. Any range of q which is of extent $2\pi/a$ contains all possible modes, but it is usual to label the modes only by values of q between $-\pi/a$ and $+\pi/a$.

The equivalence of different values of q in representing the displacement of discrete atoms is illustrated for a linear chain in Fig. 4.2. If this is to represent longitudinal vibrations, the ordinate should be considered as the longitudinal displacement of atoms which have the rest positions denoted by the open circles along the abscissa. The full circles represent the magnitude and sign of the displacements at some instant. The displacement pattern, which only has meaning at the atomic positions, can be represented as arising from any one of an infinitely large number of

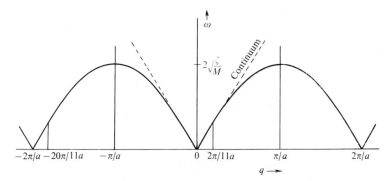

FIG. 4.1. The relation between frequency ω and wave-number, q, for vibrations of a linear chain of identical atoms with equilibrium positions separated by a distance a. The atoms are of mass M and the only forces are between nearest neighbours, the harmonic force constant being ζ. The wave numbers of the two waves of Fig. 4.2 are indicated. The broken line represents the $\omega - q$ relation for a continuum with constant velocity.

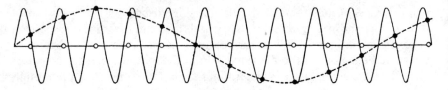

FIG. 4.2. The representation of a single pattern of atomic vibrations by two waves with very different wavelengths. (From Ziman 1960.)

different sinusoidal waves, two of which are shown. If the first particle on the left is moving towards its equilibrium position at the instant illustrated, then the longer wave is moving to the right and the shorter wave to the left. Thus $q_1 = 2\pi/\lambda_1 = +2\pi/11a$ and $q_2 = 2\pi/\lambda_2 = -2\pi(10/11a)$, where λ is the wavelength, so that $q_1 - q_2 = 2\pi/a$. These two values of q are indicated in Fig. 4.1. Energy is transported at the group velocity $d\omega/dq$, the slope of the dispersion curve, and this is the same for q_1 and q_2, as is the frequency.

The dispersion curve for a linear chain composed of different types of atoms is more complicated. If, for example, alternate atoms have different masses, half the modes are restricted to high frequencies even for small q when neighbouring atoms vibrate out of phase. The $\omega-q$ relation is shown in Fig. 4.3. The group velocity of the upper 'optic' modes is small, so that they are not very effective in transporting energy, but they may affect heat flow by interacting with the 'acoustic' modes which are mainly responsible for the conductivity.

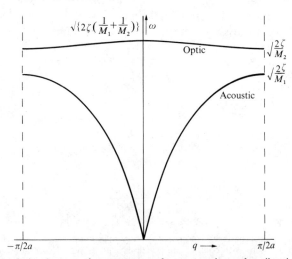

FIG. 4.3. The relation between frequency ω and wave-number q for vibrations of a linear chain in which the masses of the atoms alternate between M_1 and M_2 (for the diagram it is assumed that $M_1 > M_2$) and the nearest-neighbour harmonic force constant is ζ.

For a three-dimensional lattice of identical atoms with only one atom per 'unit cell', there are three ω–q curves corresponding to three possible polarization directions (for certain directions of propagation two of these curves, corresponding to transverse polarization, may coincide). The total number of really different modes is $3N$ if there are N atoms in the crystal, corresponding to the three degrees of vibrational freedom for each atom. This is the number of modes imposed in the Debye theory which treats a solid as a continuum. Again, if the atoms are not all the same, or if because of the lattice structure there is more than one atom per unit cell, the modes will be split among optic modes, generally of relatively high frequency, and acoustic modes with frequencies ranging from zero upwards.

For a simple cubic lattice with lattice constant a, the modes are represented by vectors **q**, with components q_x, q_y, q_z parallel to the principal axes and each lying within a range $2\pi/a$. A cube in q-space with sides of length $2\pi/a$ thus contains all possible values of **q** and its components. Such a cube, or the equivalent shape for other lattice structures, is the first Brillouin zone, and two-dimensional cross-sections of simple zones are used in Fig. 5.1 to illustrate some features of interactions among phonons. The three basic vectors, of length $2\pi/a$ for the simple cubic lattice, are the reciprocal lattice vectors **g**.

The derivation of the dispersion relations for linear chains and the general rules for constructing Brillouin zones for different lattices are given in standard books on solid state physics such as Kittel (1971).

4.1.2. Quantization of vibrational energy

According to quantum mechanics the energy in a mode of frequency ω can only have values $(n+\tfrac{1}{2})\hbar\omega$, where n is any positive integer and $2\pi\hbar$ is Planck's constant. If, as always happens in real crystals, energy can be exchanged among the different modes, statistical mechanics allows us to calculate the average value $\mathcal{N}^0$ of n in thermal equilibrium:

$$\mathcal{N}^0 = \frac{1}{\exp(\hbar\omega/k_B T) - 1}. \tag{4.2}$$

The term $\tfrac{1}{2}\hbar\omega$ is the zero-point energy, which does not vary with temperature, and the average number of thermally excited quanta of energy $\hbar\omega$ is $\mathcal{N}^0$. The quanta of vibrational energy are called phonons by analogy with the photons of electromagnetic theory.

A true normal mode of vibration and a phonon, which is its energy quantum, are spread uniformly throughout a crystal. If, however, there is a temperature gradient in a crystal which has finite conductivitity, there must be interactions which attenuate the vibrational energy, and the atomic motions do not then correspond exactly to normal modes. The thermal energy is considered as propagating by means of wave packets composed of a range of almost normal modes, which are well localized and travel with the phonon group velocity $v_G = d\omega/dq$. The attenuation is taken into account by letting the number of phonons vary with position. The value of $\mathcal{N}(q)$ in a particular region is then the number of quanta in the mode q which forms part of the wave packet in the region considered. Peierls (1955) discusses the conditions for the wave packet concept to be valid and shows that it is so for practical situations.

When heat flows through a crystal under a temperature gradient we may imagine that phonons are injected at the hot end and extracted from the cold end. It is sometimes stated that there is an analogy between this flow of phonons in a crystal under a temperature gradient and the flow of gas through a tube under a pressure gradient. Although the analogue of Knudsen low-pressure gas flow is found in all fairly good single crystals at sufficiently low temperatures, the analogue of Poiseuille flow has only been observed in a very few extremely perfect crystals, and then occurs only over a very narrow temperature range. In all other cases the interactions among the phonons which cause resistance to heat flow are not analogous to the interactions between gas molecules which, in conjunction with the walls of the tube, determine gas flow at normal pressures.

4.2. The Boltzmann equation

The heat current due to a mode q is given by the product of the thermal energy in that mode and the group velocity of propagation. The total heat current **h** carried by all modes is thus

$$\mathbf{h} = \Sigma \mathcal{N}(q) \hbar \omega \mathbf{v}_G(q). \tag{4.3}$$

In thermal equilibrium $\mathcal{N}^0(+q) = \mathcal{N}^0(-q)$ and depends only on ω. Also, $\omega(q) = \omega(-q)$ and the group velocities are equal and opposite for $+q$ and $-q$. In thermal equilibrium, therefore, $\mathbf{h} = 0$ and there will only be a net flow of heat if $\mathcal{N}(q)$ departs from the thermal equilibrium distribution in such a way that the contributions of $+q$ and $-q$ to **h** do not cancel out.

The phonon distribution numbers $\mathcal{N}(q)$ depart from equilibrium in the presence of a temperature gradient, and in the absence of any interactions would vary with time at any point in the crystal. Suppose phonons only travel in one direction, parallel to the z-axis, when there is a temperature gradient in that direction, and that at time t the number of phonons in some region is $\mathcal{N}(q)$. After a further time δt these phonons have moved away and are replaced by the density appropriate to the region $v_G^z \delta t$ away, which is

$$\mathcal{N}(q) - v_G^z \delta t \frac{\partial \mathcal{N}(q)}{\partial z}.$$

The rate of change of phonon numbers is thus

$$\left(\frac{\partial \mathcal{N}}{\partial t}\right)_{\text{drift}} = -v_G^z \frac{\partial \mathcal{N}}{\partial z} = -v_G^z \frac{\partial \mathcal{N}}{\partial T}\frac{\partial T}{\partial z}. \quad (4.4a)$$

Here q and the polarization are omitted to avoid complication, but it is obvious that the relations must be satisfied for each mode. In three dimensions the corresponding equation is

$$\left(\frac{\partial \mathcal{N}}{\partial t}\right)_{\text{drift}} = -(\mathbf{v}_G \cdot \nabla T)\frac{\partial \mathcal{N}}{\partial T}. \quad (4.4b)$$

When a steady state has been established, the phonon density at all points in the crystal becomes independent of time. There must, therefore, be processes taking place which oppose the density change due to the drift of phonons. If we refer to these as scattering processes, the Boltzmann equation can be written

$$\left(\frac{\partial \mathcal{N}}{\partial t}\right)_{\text{drift}} + \left(\frac{\partial \mathcal{N}}{\partial t}\right)_{\text{scatt.}} = 0. \quad (4.5)$$

It is in general difficult to solve this equation exactly in order to find $\mathcal{N}(q)$ from which the heat flow can be computed by eqn (4.3). For each q we have to consider scattering which takes energy from mode q and also scattering which takes energy into mode q. Apart from the problem of calculating how a particular scattering process depends on q, the rate also depends on the number density of phonons in all other modes, which is of course just what we hope to find by solving the Boltzmann equation. Methods of solution of this equation are discussed by Ziman (1960) and here the nature of only two of these methods will be indicated. Other methods which have been developed over the last 10 years or so are conceptually more difficult and do not lend themselves so easily to the analysis of experimental data. They are not discussed in this book, but a general review of them is given by Beck et al. (1974).

4.2.1. The relaxation-time method

In the relaxation time method it is assumed that scattering processes tend to restore a phonon distribution to the thermal equilibrium distribution at a rate proportional to the departure of the distribution from equilibrium, so that

$$\left(\frac{\partial \mathcal{N}}{\partial t}\right)_{\text{scatt.}} = \frac{\mathcal{N}^0 - \mathcal{N}}{\tau} \quad (4.6)$$

where the relaxation time τ as well as $\mathcal{N}^0$ and $\mathcal{N}$, depends on q and on polarization. It is usually further assumed that the distribution in the presence of a temperature gradient is not too greatly different from the equilibrium distribution, so that in the drift term of the Boltzmann equation (4.5) we can replace $\partial \mathcal{N}/\partial T$ by $\partial \mathcal{N}^0/\partial T$. From eqns (4.4a), (4.5), and (4.6) we obtain for one dimension

$$v_G^z \frac{\partial \mathcal{N}^0}{\partial T} \frac{\partial T}{\partial z} = \frac{\mathcal{N}^0 - \mathcal{N}}{\tau}. \qquad (4.7)$$

This equation gives directly the deviation from the thermal equilibrium distribution which, as we have seen, gives rise to a flow of heat. Combining eqns (4.3) and (4.7) yields

$$h = -\sum \hbar \omega (v_G^z)^2 \tau \frac{\partial \mathcal{N}^0}{\partial T} \frac{\partial T}{\partial z}$$

and the heat conductivity

$$\kappa = -\frac{h}{(\partial T/\partial z)} = \sum \hbar \omega (v_G^z)^2 \tau \frac{\partial \mathcal{N}^0}{\partial T}.$$

If we replace the summation by an integral over ω and write $(v_G^z)^2 = \tfrac{1}{3} v_G^2$, we obtain

$$\kappa = \tfrac{1}{3} \int_0^{\omega_{max}} \hbar \omega v_G^2 \tau \mathrm{f}(\omega) \frac{\partial \mathcal{N}^0}{\partial T} d\omega \qquad (4.8)$$

where $\mathrm{f}(\omega)\, d\omega$ is the number of phonon modes between ω and $\omega + d\omega$ per unit volume of crystal.

As will become clear, the actual computation of thermal conductivities is so fraught with other difficulties that there is little point in trying to evaluate eqn (4.8) for the precise frequency spectrum and dispersion curve of a real crystal. We lose little if at this stage we take over some of the simplifying assumptions of the Debye theory in order to reduce eqn (4.8) to a fairly manageable form. The Debye model assumes a simple linear dispersion relation of the form $\omega(q) = vq$ for each branch of the phonon spectrum (this is the limiting form for small q in real crystals). If it is further assumed that phonon velocities are the same for the three polarizations, then the fundamental relation

$$\mathrm{f}(q)\, dq = \frac{3q^2\, dq}{2\pi^2}$$

leads to

$$\mathrm{f}(\omega)\, d\omega = \frac{3}{2\pi^2 v^3} \omega^2\, d\omega.$$

From eqn (4.2)

$$\frac{\partial \mathcal{N}^0(\omega)}{\partial T} = \frac{(\hbar\omega/k_B T^2) \exp(\hbar\omega/k_B T)}{\{\exp(\hbar\omega/k_B T) - 1\}^2}$$

so that

$$\kappa = \frac{1}{2\pi^2 v} \int_0^{\omega_{max}} \hbar\omega^3 \tau \frac{(\hbar\omega/k_B T^2) \exp(\hbar\omega/k_B T)}{\{\exp(\hbar\omega/k_B T) - 1\}^2} d\omega. \quad (4.9a)$$

If we make the standard substitution $x = \hbar\omega/k_B T$ and write the maximum frequency $\omega_{max} = \theta(k_B/\hbar)$, where the Debye temperature θ is chosen to be such that there are just $3N$ modes, eqn (4.9a) becomes

$$\kappa = \frac{k_B}{2\pi^2 v} \left(\frac{k_B}{\hbar}\right)^3 T^3 \int_0^{\theta/T} \tau(x) \frac{x^4 e^x}{(e^x - 1)^2} dx. \quad (4.9b)$$

The contribution to the heat capacity from modes in the range ω to $\omega + d\omega$ is $(d/dT)\{\hbar\omega \mathcal{N}^0(\omega) f(\omega) d\omega\}$. In terms of the dimensionless parameter x the differential contribution to the heat capacity is, in the Debye approximation,

$$C(x) dx = \frac{3k_B}{2\pi^2 v^3} \left(\frac{k_B}{\hbar}\right)^3 T^3 \frac{x^4 e^x}{(e^x - 1)^2} dx \quad (4.10)$$

and the conductivity given by eqn (4.9b) can be written as

$$\kappa = \tfrac{1}{3} v^2 \int_0^{\theta/T} \tau(x) C(x) dx. \quad (4.11a)$$

A relaxation time can be expressed as the ratio of a mean free path to a velocity, so that the conductivity may also be written as

$$\kappa = \tfrac{1}{3} v \int l(x) C(x) dx \quad (4.11b)$$

which is a logical extension of the simple kinetic equation already used (eqn (3.5)).

It would seem at first sight that the problem is to compute the $\tau_i(x)$ appropriate to each scattering mechanism i and add the scattering rates $\tau_i^{-1}(x)$ to obtain $\tau(x)$ for insertion under the integral of eqn (4.11a). Although there are circumstances when this method is adequate to explain experimentally measured conductivities and their dependence on the concentration of scatterers, such as lattice defects, the situation is generally more complicated. Further discussion of this point must wait until phonon interactions have been discussed; it will be seen (§ 5.2) that there are interactions between phonons which do not tend to restore an arbitrary distribution to the thermal equilibrium distribution in the manner envisaged in eqn (4.6) and yet cannot be ignored.

4.2.2. The variational method

This method, which had been used earlier for electrical conductivity, was first applied to lattice thermal conduction by Leibfried and Schlömann (1954) and by Ziman (1956a, 1960).

In the relaxation-time method one equates, for the steady state, the rate of change of the phonon distribution due to phonon drift in the presence of a temperature gradient with the rate of change due to scattering processes. For each q one assumes an exponential relaxation to equilibrium due to the scattering, which is independent of departures from equilibrium of all the other modes even though interactions involving these other modes are concerned in the relaxation to equilibrium.

Although the nature of the phonon distribution does not appear directly in an expression such as eqn (4.8) for the conductivity, the distribution which corresponds to any particular form for $\tau(q, T)$ can be deduced from eqn (4.7). In the variational method, on the other hand, one is directly concerned with the phonon distribution and must seek that form of it which satisfies a variational criterion which is equivalent to minimizing the thermal resistivity.

The rate of change in the density $\mathcal{N}$ of phonons of a particular mode depends on the intrinsic probabilities of the scattering processes involved and on the densities of all the interacting modes (the population factor). For any given phonon distribution there will thus be a thermal resistivity corresponding to a particular mechanism or combination of mechanisms. According to the variational principle it is necessary to deduce that form for the distribution which will yield the minimum resistivity when the scattering processes take place in a phonon system so distributed. For some scattering processes acting separately it is relatively simple to find the corresponding phonon distributions such that the resistivity is zero, and this is certainly the minimum possible (the case of scattering by point defects is discussed here and that of phonon–phonon normal processes in § 5.2), but, if different types of process occur together, the correct form for the distribution is in general difficult to find (see § 6.2.1).

Ziman has discussed how the variational method is equivalent to equating the rate of production of entropy by the scattering processes to its rate of production due to heat flowing down the temperature gradient. The former can be expressed in terms of the rate of change in $\mathcal{N}$ and a function describing the departure of the distribution from equilibrium. In order to obtain the latter, an entropy flux $\mathbf{h}/T$ is associated with the heat flux $\mathbf{h}$. The rate of entropy production can then be written

$$\left(\frac{\partial \mathscr{S}}{\partial t}\right)_{\text{flow}} = \mathbf{h}\cdot\nabla\left(\frac{1}{T}\right) = -\frac{\nabla T\cdot\mathbf{h}}{T^2} = \frac{h^2}{\kappa T^2}.$$

The displaced phonon distribution is written

$$\mathcal{N}(\mathbf{q}) = \mathcal{N}^0(\mathbf{q}) - \phi(\mathbf{q})\frac{\partial \mathcal{N}^0(\mathbf{q})}{\partial E(\mathbf{q})} \qquad (4.12)$$

where $\phi(\mathbf{q})$ is a function measuring the deviation of $\mathcal{N}(\mathbf{q})$ for phonons $\mathbf{q}$

PHONONS AND THE BOLTZMANN EQUATION

from the equilibrium distribution $\mathcal{N}^0(\mathbf{q})$. Since

$$\mathcal{N}^0(\mathbf{q}) = \frac{1}{\exp\{E(\mathbf{q})/k_B T\} - 1}$$

$$\frac{\partial \mathcal{N}^0(\mathbf{q})}{\partial E(\mathbf{q})} = -\frac{1}{k_B T} \frac{\exp(E/k_B T)}{\{\exp(E/k_B T) - 1\}^2}$$

$$= -\frac{1}{k_B T} \mathcal{N}^0(\mathbf{q})\{1 + \mathcal{N}^0(\mathbf{q})\}$$

and

$$\mathcal{N}(\mathbf{q}) = \mathcal{N}^0(\mathbf{q}) + \phi(\mathbf{q}) \frac{\mathcal{N}^0(\mathbf{q})\{1 + \mathcal{N}^0(\mathbf{q})\}}{k_B T}. \tag{4.13}$$

As was pointed out before, a heat flow is only produced by departures of the distribution from equilibrium, so that $\mathbf{h}$ can be expressed as

$$\mathbf{h} = -\sum \hbar \omega \mathbf{v}_G(\mathbf{q}) \phi(\mathbf{q}) \frac{\partial \mathcal{N}^0(\mathbf{q})}{\partial E(\mathbf{q})}. \tag{4.14}$$

In order to see the form of the expression which has to be minimized, we consider elastic scattering of phonons by a defect, which in an isotropic material results in a phonon $\mathbf{q}$ being scattered into a mode $\mathbf{q}'$ with the same magnitudes of q and ω but different in direction.

The transition rate from $\mathbf{q}$ to $\mathbf{q}'$ is proportional to the population factor $\mathcal{N}(\mathbf{q})\{1 + \mathcal{N}(\mathbf{q}')\}$ and to the scattering probability $Q(\mathbf{q}, \mathbf{q}')$, with a similar expression for the reverse transition. The rate of change of the number of phonons in mode $\mathbf{q}$ is thus

$$\left(\frac{\partial \mathcal{N}}{\partial t}\right)_{\text{scatt.}} = \sum_{\mathbf{q}'} [\mathcal{N}(\mathbf{q}')\{1 + \mathcal{N}(\mathbf{q})\}Q(\mathbf{q}', \mathbf{q}) - \mathcal{N}(\mathbf{q})\{1 + \mathcal{N}(\mathbf{q}')\}Q(\mathbf{q}, \mathbf{q}')].$$

From the principle of microscopic reversibility, $Q(\mathbf{q}, \mathbf{q}') = Q(\mathbf{q}', \mathbf{q})$ so that

$$\left(\frac{\partial \mathcal{N}}{\partial t}\right)_{\text{scatt.}} = \sum_{\mathbf{q}'} [\{\mathcal{N}(\mathbf{q}') - \mathcal{N}(\mathbf{q})\}Q(\mathbf{q}, \mathbf{q}')].$$

Writing

$$P(\mathbf{q}, \mathbf{q}') = \mathcal{N}^0(\mathbf{q})\{1 + \mathcal{N}^0(\mathbf{q})\}Q(\mathbf{q}, \mathbf{q}')$$

and noting that $\mathcal{N}^0(\mathbf{q}) = \mathcal{N}^0(\mathbf{q}')$, since we are assuming that $\mathbf{q}$ and $\mathbf{q}'$ correspond to the same energy,

$$\left(\frac{\partial \mathcal{N}}{\partial t}\right)_{\text{scatt.}} = \frac{1}{k_B T} \sum_{\mathbf{q}'} \{\phi(\mathbf{q}') - \phi(\mathbf{q})\} P(\mathbf{q}, \mathbf{q}').$$

The rate of entropy production by the scattering processes is derived from the statistical definition of entropy:

$$\mathscr{S} = -k_B \sum_{\mathbf{q}} [\mathscr{N}(\mathbf{q}) \ln \mathscr{N}(\mathbf{q}) - (1 + \mathscr{N}(\mathbf{q})) \ln \{1 + \mathscr{N}(\mathbf{q})\}].$$

Differentiation with respect to time and substitution of the expression for $\mathscr{N}(\mathbf{q})$ (eqn (4.13)) gives

$$\left(\frac{\partial \mathscr{S}}{\partial t}\right)_{\text{scatt.}} = -k_B \sum \left(\frac{\partial \mathscr{N}}{\partial t}\right)_{\text{scatt.}} \left[\ln\left\{1 + \frac{\phi \exp(E/k_B T)}{k_B T}\mathscr{N}^0\right\} \right. $$
$$\left. - \ln\left\{\exp(E/k_B T)\left(1 + \frac{\phi \mathscr{N}^0}{k_B T}\right)\right\}\right]$$

where the dependences of $\mathscr{N}$, ϕ, and E on $\mathbf{q}$ are omitted. If the distribution is near equilibrium, the logarithms can be expanded to only the first power of ϕ, so that

$$\left(\frac{\partial \mathscr{S}}{\partial t}\right)_{\text{scatt.}} = \sum \left(\frac{\partial \mathscr{N}}{\partial t}\right)_{\text{scatt.}} \left[\frac{E}{T} - \frac{\phi}{T}\{\exp(E/k_B T) - 1\}\mathscr{N}^0\right]$$
$$= \sum \left\{\frac{E(\mathbf{q})}{T}\frac{\partial \mathscr{N}(\mathbf{q})}{\partial t} - \frac{\phi(\mathbf{q})}{T}\frac{\partial \mathscr{N}(\mathbf{q})}{\partial t}\right\}.$$

The first term is zero for the steady state in which the total energy of the system is constant, and

$$\left(\frac{\partial \mathscr{S}}{\partial t}\right)_{\text{scatt.}} = -\frac{1}{T}\sum_{\mathbf{q}} \phi(\mathbf{q})\left(\frac{\partial \mathscr{N}(\mathbf{q})}{\partial t}\right)_{\text{scatt.}}$$

which for our example becomes

$$\frac{1}{k_B T^2} \sum_{\mathbf{q}} \sum_{\mathbf{q}'} \phi(\mathbf{q})\{\phi(\mathbf{q}) - \phi(\mathbf{q}')\} P(\mathbf{q}, \mathbf{q}').$$

Because the summation is symmetrical in $\mathbf{q}$ and $\mathbf{q}'$ this expression may be written

$$\frac{1}{2k_B T^2} \sum_{\mathbf{q}} \sum_{\mathbf{q}'} \{\phi(\mathbf{q}) - \phi(\mathbf{q}')\}^2 P(\mathbf{q}, \mathbf{q}').$$

Equating this expression for $\partial \mathscr{S}/\partial t$ to the value $h^2/\kappa T^2$ arising from the flow of heat, we obtain

$$\frac{1}{\kappa} = \frac{(1/2k_B T^2) \sum_{\mathbf{q}} \sum_{\mathbf{q}'} \{\phi(\mathbf{q}) - \phi(\mathbf{q}')\}^2 P(\mathbf{q}, \mathbf{q}')}{\left[(1/T) \sum_{\mathbf{q}} \hbar\omega(\mathbf{q}) v_G(\mathbf{q}) \phi(\mathbf{q}) \{\partial \mathscr{N}^0(\mathbf{q})/\partial E(\mathbf{q})\}\right]^2}. \quad (4.15)$$

This is the expression which must be minimized by choosing $\phi(\mathbf{q})$ appropriately.

In order that there should be a net heat flow down the temperature gradient, the phonon distribution must be asymmetrical with respect to the direction of grad T. A suitable form of trial function for the type of scattering process being discussed is

$$\phi(q) = \frac{1}{q^n}(\mathbf{q} \cdot \mathbf{u}),$$

where n is not necessarily an integer and $\mathbf{u}$ is an arbitrary vector in the direction of the temperature gradient (its absolute value is determined by the magnitude of the heat flow but does not enter into the expression for $1/\kappa$ because it appears squared in both the numerator and denominator).

If the resistance is due to point defects, scattering according to the Rayleigh law ($Q(\mathbf{q}', \mathbf{q}) \propto q^2$), there are in fact an infinite number of forms of $\phi(q)$ which give the same minimum value $1/\kappa = 0$. For $n \geq 3.5$, both the numerator and denominator of eqn (4.15) diverge as $q \to 0$, but maintain a ratio equal to zero. Sheard (private communication) has shown how the correct unique form of the phonon distribution can, nevertheless, be deduced for Rayleigh scattering. If it is assumed to start with that the phonon spectrum does not extend down quite to $q = 0$, but stops at $q = q_{min}$, then eqn (4.15) leads to an expression for $1/\kappa$ proportional to $q_{min}\{(n-3)^2/(2n-7)\}$. If the resistance is to be positive, then $n \geq 3.5$ and the minimum occurs for $n = 4$. If q_{min} is now reduced indefinitely, $1/\kappa$ is still a minimum for $n = 4$ and this minimum is zero for $q_{min} = 0$.

This is certainly the minimum value which we can hope to find and must therefore be the thermal resistivity which would result if point defects were the only source of scattering. The phonons take up the non-equilibrium distribution given by $\phi(\mathbf{q}) = (1/q^4)(\mathbf{q} \cdot \mathbf{u})$.

The relaxation time expression (4.9b) also gives a resistivity equal to zero if scattering by point defects is the only kind of scattering to be considered. If we take $\tau(q) \propto q^{-4}$, then for small q (so that $\omega \propto q$), $\tau(x) \propto x^{-4}T^{-4}$, and the variables under the integral become $\exp(x)/\{\exp(x)-1\}^2$. The integral diverges for small x and the heat conductivity is infinite.

We can also compare the phonon distributions for point-defect scattering deduced from both the variational and relaxation-time methods. The departure from equilibrium represented by eqn (4.13), using the variational solution $\phi(q) = (1/q^4)(\mathbf{q} \cdot \mathbf{u})$ is

$$\mathcal{N}(\mathbf{q}) = \mathcal{N}^0(\mathbf{q}) + \frac{1}{q^4}(\mathbf{q} \cdot \mathbf{u})\frac{\mathcal{N}^0(\mathbf{q})(1+\mathcal{N}^0(\mathbf{q}))}{k_B T}.$$

For the relaxation-time method, eqn (4.7) or its three-dimensional analogue is used. Substituting $\tau(q) \propto q^{-4}$ we obtain

$$\mathcal{N}^0(\mathbf{q}) - \mathcal{N}(\mathbf{q}) \propto q^{-4}\mathbf{v} \cdot \frac{\partial \mathcal{N}^0(\mathbf{q})}{\partial T} \nabla T.$$

Let us assume, for simplicity, that there is no dispersion, so that $\partial \mathcal{N}^0(\mathbf{q})/\partial T$ is proportional to $q\mathcal{N}^0(\mathbf{q})\{1+\mathcal{N}^0(\mathbf{q})\}$, and that $\mathbf{v}$ is in the same direction as $\mathbf{q}$, so that $q\mathbf{v}$ is proportional to $\mathbf{q}$; then

$$\mathcal{N}^0(\mathbf{q}) - \mathcal{N}(\mathbf{q}) \propto \frac{1}{q^4}(\mathbf{q}\cdot\nabla T)\mathcal{N}^0(\mathbf{q})\{1+\mathcal{N}^0(\mathbf{q})\}.$$

Both ∇T and $\mathbf{u}$ are vectors in the direction of the temperature gradient, so that apart from an apparent numerical factor, the two expressions for $\mathcal{N} - \mathcal{N}^0$ agree. If the two expressions are made to refer to the same heat flow, they are even numerically the same.

For real situations in which several types of scattering processes act together, the simple relaxation-time method and the variational method can give very different results. However, the two methods are brought into good agreement again when Callaway's (1959) modification of the relaxation time method is used to take account of those interactions among phonons which do not by themselves tend to restore an arbitrary distribution to the thermal equilibrium distribution.

5
NORMAL AND UMKLAPP PROCESSES

IF THE potential energy of an atom in a perfect crystal were exactly quadratic in the displacement from equilibrium, thermal waves would propagate without interacting with one another. There would be no tendency for an arbitrary phonon distribution to return to the equilibrium distribution, even in the absence of a temperature gradient. Since there would be no mechanism for equalizing $\mathcal{N}(+q)$ and $\mathcal{N}(-q)$, the heat flow given by eqn (4.3) would continue indefinitely and the conductivity would be infinite. In the relaxation-time expression (4.9b) for the conductivity, $\tau(x)$ for all modes would be infinite, and similarly in the variational expression for the resistivity, eqn (4.15), the transition probability P in the numerator would be zero.

Experience suggests that if a large defect-free crystal could be produced its conductivity would, nevertheless, be finite. Even in an ideal crystal the phonon lifetimes are limited by processes which arise because the potential is not strictly quadratic in the atomic displacements. We now consider the nature of these interactions.

5.1. Phonon–phonon interactions

If the particles of an anharmonic medium are displaced by a wave motion, then the elastic properties as seen by another wave will vary along the path of the first wave. Where the wave has displaced the particles further the medium will be stiffer (or less stiff, depending on the sign of the departure from harmonicity) than where the displacements are small. This idea formed the basis of Debye's theory (1914) of thermal conductivity. He treated the effect of all the waves except the one being considered as represented by static regions of differing elastic properties in a continuous medium, and obtained an expression for the rate of attenuation. In fact, the waves which combine to form the regions of differing elastic properties propagate at the same velocity as the wave being scattered, and a proper calculation shows that the mechanism envisaged would not lead to a finite thermal resistance in a continuous medium. It is only the treatment of a crystal as consisting of a discrete lattice that leads to the result that even a perfect crystal has a finite conductivity.

In order to find the conditions under which modes of vibration gain or lose energy to other modes, we can use time-dependent perturbation theory in which the perturbing Hamiltonian arises from that part of the potential energy of a displaced atom which is associated with powers of the displacement higher than the square (the squared term is normally sufficient to characterize the vibrational energy and specific heat). There may be cubic, quartic, or higher terms, but we shall concentrate on the cubic term, which is most important for the conductivity, except possibly at very high temperatures, and illustrates in principle the possible kinds of effect which phonon–phonon interactions can have on the thermal conductivity.

The probability of the number of phonons in a mode $(\mathbf{q}_1, \omega_1)$, characterized by wave vector $\mathbf{q}_1$ and frequency ω_1, changing due to the cubic anharmonicity is expressed in terms of processes which involve interactions among three modes; for example the energy in modes $(\mathbf{q}_1, \omega_1)$ and $(\mathbf{q}_2, \omega_2)$ may appear after the interaction in mode $(\mathbf{q}_3, \omega_3)$. This process may take place in the opposite direction or the energy in mode $(\mathbf{q}_1, \omega_1)$ may be divided among two other phonons, the reverse process again being possible. The quartic contribution to the potential energy leads to interactions involving four phonons. The structure of the expression for the transition probability for three-phonon processes is such that it is effectively zero unless two conditions are fulfilled:

$$\omega_1 + \omega_2 = \omega_3 \qquad (5.1a)$$

and

$$\mathbf{q}_1 + \mathbf{q}_2 = \mathbf{q}_3 + \mathbf{g} \qquad (5.1b)$$

where $\mathbf{g}$ is a reciprocal lattice vector or 0.

Rather than discuss the derivation of these two equations, which was first given by Peierls (1929) (see also Peierls 1955), we shall concentrate here on their interpretation and on their consequences for thermal conductivity.

5.1.1. Normal and Umklapp processes

Since $\hbar\omega$ is the quantum of energy for a mode of frequency ω, eqn (5.1a) represents the conservation of energy in a three-phonon process. A vibrational mode does not have a mechanical momentum as does a material particle, but $\hbar\mathbf{q}$ does have some of the attributes of momentum. Eqn (5.1b) with $\mathbf{g} = 0$ thus looks like a statement of the law of conservation of momentum. An interaction for which $\mathbf{g} = 0$ is now called a normal process, and Peierls called processes for which $\mathbf{g} \neq 0$ an Umklapp process. They will be referred to as N- and U-processes respectively.

The difference between N- and U-processes can be illustrated by drawing a two-dimensional cross-section of the Brillouin zone discussed in § 4.1.1. With the centre as origin the two vectors $\mathbf{q}_1$ and $\mathbf{q}_2$ are drawn and their vector sum represents their resultant $\mathbf{q}_3$, as shown in Fig. 5.1a. In Fig. 5.1b (and in Fig. 5.1c, to be discussed in § 7.1.2(b)) the initial

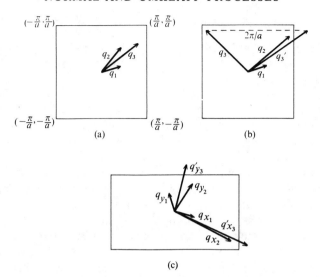

FIG. 5.1. Two-dimensional representation of three-phonon processes: (a) resultant q_3 falls within the Brillouin zone—normal process; (b) resultant q_3' falls outside the Brillouin zone—Umklapp process; (c) for a Brillouin zone represented by a rectangle the minimum value of q_3' along one axis, corresponding to a U-process, is different from that along the other axis.

vectors have been chosen so that their resultant falls beyond the edge of the zone and has been labelled $\mathbf{q}_3'$. In the one-dimensional case of a linear chain it was shown that modes of vibration for which the values of q differ by $2\pi/a$ represent the same pattern of atomic motions. Similarly in the three-dimensional case, modes which are represented by values of $\mathbf{q}$ which differ by a reciprocal lattice vector represent the same atomic motions, and the mode $\mathbf{q}_3'$ is identical with the mode $\mathbf{q}_3$ obtained by adding or subtracting $2\pi/a$ (for our simple cubic lattice) to the vector $\mathbf{q}_3'$. In a U-process, therefore, the vector sums on the two sides of eqn (5.1b) have to differ by $\mathbf{g}$ if all the phonons are to be represented by vectors lying inside the first Brillouin zone.

Thermal energy is carried in the direction of the phonon group velocity, and in an N-process the direction of energy flow carried by mode $\mathbf{q}_3$ is evidently similar to the effective direction of the energy flow carried by $\mathbf{q}_1$ and $\mathbf{q}_2$. It will, indeed, be shown in the next section that if the only interactions undergone by phonons are N-processes then a crystal would have infinite thermal conductivity. After a U-process thermal energy is transported in quite a different direction, that of the group velocity of $\mathbf{q}_3$. The action of such drastic changes in $\mathbf{q}$ results in a tendency to return any phonon distribution to the equilibrium form.

5.2. The effect of N-processes

Before discussing the subtle influence of N-processes on thermal conductivity when other processes occur, we shall show that by themselves they cannot maintain a finite conductivity.

We may first make this idea plausible by analogy with the flow of gas in a tube. On the average, the collisions of molecules with one another may be considered to conserve energy and momentum, and are thus analogous to N-processes among phonons. When gas at normal pressure flows through a tube, the molecules are constantly colliding with one another and a well-defined distribution of velocities is set up, consistent with the net drift velocity observed. In a real situation this distribution varies over the cross-section of the tube since the drift velocity varies with the distance from the axis of the tube. If the walls of the tube were infinitely far away or completely smooth, so that only specular reflection of molecules occurred on impact with them, or if the gas were contained in a box which slid through the tube without friction, then although the molecules would still be colliding with one another there would be no resistance to gas flowing through the tube. Under these conditions the molecules would have a definite velocity distribution, which would be different from the equilibrium Maxwell–Boltzmann distribution corresponding to no net flow but would remain unchanged under the action of molecular collisions.

In the same way, if heat flows along an infinitely large perfect crystal, or one which has walls which reflect phonons specularly, the energy and 'momentum'-conserving N-processes will not affect the distribution which corresponds to the net heat flow. This distribution is different from the equilibrium Planck distribution given by eqn (4.2). If there can be a phonon distribution corresponding to a heat flow which is not changed by N-processes, then N-processes alone will not give rise to thermal resistance.

We can make the deduction of the ineffectiveness of N-processes from Fig. 5.1a more precise if we assume that all phonons have the same velocity and that this is parallel to $\mathbf{q}$. Then $\omega = qv$ and eqn (4.3) becomes $\mathbf{h} = \hbar v^2 \sum \mathcal{N}(q)\mathbf{q}$. In an N-process one phonon is removed from each of the modes $\mathbf{q}_1$ and $\mathbf{q}_2$ so that $\sum \mathcal{N}(q)\mathbf{q}$ is reduced by $\mathbf{q}_1 + \mathbf{q}_2$. After the process $\sum \mathcal{N}(q)\mathbf{q}$ has been increased by $\mathbf{q}_3$, the net change in $\mathbf{h}$ is thus

$$\hbar v^2 (\mathbf{q}_3 - \mathbf{q}_1 - \mathbf{q}_2) = 0.$$

The heat flow is therefore unchanged by an N-process.

This simple proof relies on the absence of dispersion in the $\omega - q$ relation, but the result is, in fact, quite general. One way of showing this is to examine the behaviour under the action of N-processes alone of the 'displaced phonon distribution' given by

$$\mathcal{N}(\mathbf{q}, \omega) = \frac{1}{\exp\{\hbar(\omega - \mathbf{q}\cdot\mathbf{u})/k_B T\} - 1}. \quad (5.2)$$

In this expression the vector $\mathbf{u}$ is in the direction of heat flow and represents a net drift velocity. The distribution is of the equilibrium form relative to a coordinate system which moves with velocity $\mathbf{u}$. A one-dimensional representation of such a distribution is shown in Fig. 5.2, and it is obvious from its asymmetry that it corresponds to a net heat flow to the right.

The perturbation theory expression for the rate of change of $\mathcal{N}(q)$ due to N-processes contains the probability of a process taking place which has a phonon of this mode as an end product and also the probability of a process in which a phonon of this mode disappears as a result of interactions involving two other phonons. If we consider the process $\mathbf{q}_1 + \mathbf{q}_2 \rightarrow \mathbf{q}_3$ and the reverse process, then the rate of change of $\mathcal{N}(\mathbf{q}_1)$ contains the difference between the factors

$$\{\mathcal{N}(\mathbf{q}_1)\}\{\mathcal{N}(\mathbf{q}_2)\}\{\mathcal{N}(\mathbf{q}_3) + 1\}$$

from the forward process and

$$\{\mathcal{N}(\mathbf{q}_1) + 1\}\{\mathcal{N}(\mathbf{q}_2) + 1\}\{\mathcal{N}(\mathbf{q}_3)\}$$

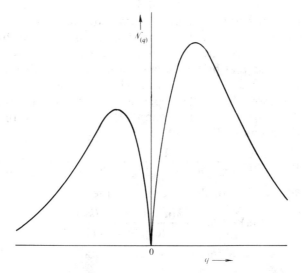

FIG. 5.2. An asymmetrical phonon distribution which is unaffected by N-processes and corresponds, in this one-dimensional representation, to a heat flow to the right. The particular distribution illustrated is given by eqn (5.2) with $u = 0\cdot1\omega/q$ and $T = 100$ K. The ordinate is the number of phonons per unit interval in q.

from the reverse process (these weighting factors are quite general for systems obeying Bose–Einstein statistics).

If we substitute the expression (5.2) for $\mathcal{N}(\mathbf{q})$ into these two factors and take their difference, a simple calculation leads to the term

$$\frac{\exp(-\hbar \mathbf{q}_3.\mathbf{u}/k_BT)\exp(\hbar\omega_3/k_BT) - \exp\{-\hbar(\mathbf{q}_1+\mathbf{q}_2).\mathbf{u}/k_BT\}\exp\{\hbar(\omega_1+\omega_2)/k_BT\}}{[\exp\{\hbar(\omega_1-\mathbf{q}_1.\mathbf{u})/k_BT\}-1][\exp\{\hbar(\omega_2-\mathbf{q}_2.\mathbf{u})/k_BT\}-1][\exp\{\hbar(\omega_3-\mathbf{q}_3.\mathbf{u})/k_BT\}-1]}$$

representing the phonon-density contribution to the rate of change of $\mathcal{N}(\mathbf{q}_1)$. If eqns (5.1) are satisfied with $\mathbf{g}=0$, so that only N-processes are considered, we see that the rate of change of $\mathcal{N}(\mathbf{q}_1)$ due to the processes considered is zero. We obtain the same result for all pairs of values of $\mathbf{q}_2$ and $\mathbf{q}_3$ which can take part in an N-process, as we do if we consider processes of the form $\mathbf{q}_1 \to \mathbf{q}_2+\mathbf{q}_3$ which also affect the number of phonons $\mathbf{q}_1$. The displaced phonon distribution given by eqn (5.2) which corresponds to a finite heat flow is thus unaffected by N-processes.

The same result can also be derived from the variational principle. For three-phonon N-processes the numerator of eqn (4.15) is modified to

$$\sum_{\mathbf{q}_1}\sum_{\mathbf{q}_2}\sum_{\mathbf{q}_3}\{\phi(\mathbf{q}_1)+\phi(\mathbf{q}_2)-\phi(\mathbf{q}_3)\}P_{\mathbf{q}_3}^{\mathbf{q}_1,\mathbf{q}_2}. \tag{5.3}$$

If we write

$$\phi(\mathbf{q}) = \mathbf{q}.\mathbf{u}' \tag{5.4}$$

where $\mathbf{u}'$ is an arbitrary vector, then the numerator contains $(\mathbf{q}_1+\mathbf{q}_2-\mathbf{q}_3).\mathbf{u}'$, which is zero if $\mathbf{q}_1+\mathbf{q}_2=\mathbf{q}_3$. By using this trial function we have found a value for the thermal resistivity equal to zero, which is certainly a minimum value, so that by the variational principle the thermal conductivity in the presence of N-processes alone is infinite.

It is easy to show that, if the departure from equilibrium is small, the expression for $\mathcal{N}(q)$ derived from eqn (5.4) is equivalent to that given in eqn (5.2). From eqn (5.2) we obtain

$$\mathcal{N}(\mathbf{q}) - \mathcal{N}^0(\mathbf{q}) = \frac{1}{\exp\{\hbar(\omega-\mathbf{q}.\mathbf{u})/k_BT\}-1} - \frac{1}{\exp(\hbar\omega/k_BT)-1}$$

and when $\hbar\mathbf{q}.\mathbf{u}/k_BT$ is small this can be written

$$\mathcal{N}(\mathbf{q}) - \mathcal{N}^0(\mathbf{q}) = \frac{\hbar(\mathbf{q}.\mathbf{u})\{\exp(\hbar\omega/k_BT)/k_BT\}}{\{\exp(\hbar\omega/k_BT)-1\}^2}.$$

Substituting $\phi(\mathbf{q})=\mathbf{q}.\mathbf{u}'$ into eqn (4.12) we obtain

$$\mathcal{N}(\mathbf{q}) - \mathcal{N}^0(\mathbf{q}) = -\mathbf{q}.\mathbf{u}'\frac{\partial \mathcal{N}^0(\mathbf{q})}{\partial E(\mathbf{q})}$$

$$= \frac{\mathbf{q}.\mathbf{u}'}{k_BT}\frac{\exp(\hbar\omega/k_BT)}{\{\exp(\hbar\omega/k_BT)-1\}^2}.$$

These two expressions are the same if $\hbar\mathbf{u} = \mathbf{u}'$. The values of $\mathbf{u}$ and $\mathbf{u}'$ are arbitrary and are determined by the magnitude of the heat flow, and for a given heat flow the two expressions for the departure from the equilibrium distribution are the same.

It must not be thought that because N-processes themselves do not give rise to thermal resistance they may be safely ignored. They can have a profound effect because the scattering rates for other processes may be frequency dependent; in this situation N-processes prevent modes which would be little scattered by such processes from 'running away' with the heat flow. Much effort has been expended in attempting to explain how N-processes do combine with resistive processes to determine thermal conductivity. This problem will be discussed in the next chapter.

6

TAKING ACCOUNT OF NORMAL PROCESSES

ALL scattering processes which phonons may undergo that tend to restore a non-equilibrium distribution to equilibrium have a direct effect on the thermal conductivity. For most processes the scattering rate depends on the phonon frequency, and N-processes can then have an important influence by transferring energy between different modes and thus preventing any from deviating much from the equilibrium population. In general, the effect of N-processes is not immediately apparent, and fairly detailed analysis of experimental results is necessary to see how the conductivity has been affected by their existence. However, there are cases in which their influence is dramatic.

Various treatments of N-processes will be discussed before individual scattering mechanisms are described in chapter 7.

6.1. The relaxation-time method

In the simplest form of the relaxation-time method it is assumed that each scattering mechanism can be characterized by relaxation times with values that, for a given mode, are independent of the phonon populations of all other modes. If several types of scattering process act together, the scattering rates for each mode, $\tau_i^{-1}(x)$, $\tau_j^{-1}(x)$, ..., are added together and the resultant relaxation time $\tau(x)$, given by

$$\tau^{-1}(x) = \sum_i \tau_i^{-1}(x),$$

is substituted into an expression such as eqn (4.9b). This expression would, of course, be more involved if we wanted to go beyond the simplifying Debye assumptions of no dispersion and a quadratic mode density, $f(\omega) \propto \omega^2$.

Since N-processes themselves do not tend to restore the equilibrium phonon distribution, they cannot come into the summation for $\tau(x)$ on the same footing as processes which do tend to restore the equilibrium distribution. On the other hand, they cannot be ignored because, by bringing about exchanges of energy between modes, they make sure that the presence of frequency-dependent resistive scattering processes is felt by all modes.

The most widely used treatment in analysing experimental thermal conductivity data is that of Callaway (1959). He assumed that N-processes restore an arbitrary phonon distribution, representing a heat flow, to the distribution (eqn (5.2)) which corresponds to the same heat flow but is no longer changed by N-processes. The relaxation time for such processes is τ_N (for simplicity a dependence of this relaxation time on q, polarization, and temperature is not indicated by further symbols). The total rate of change of $\mathcal{N}(q)$ is then expressed as

$$\frac{\partial \mathcal{N}}{\partial t} = -\frac{\mathcal{N} - \mathcal{N}^0}{\tau_R} - \frac{\mathcal{N} - \mathcal{N}(\mathbf{u})}{\tau_N}$$

where only processes which tend to restore the true thermal equilibrium distribution $\mathcal{N}^0(q)$ contribute to τ_R. Assuming a Debye model for the solid, and writing a combined relaxation rate $\tau_C^{-1} = \tau_R^{-1} + \tau_N^{-1}$, Callaway's expression for the thermal conductivity can be written

$$\kappa = \kappa_1 + \kappa_2 \qquad (6.1)$$

where

$$\kappa_1 = \frac{k_B}{2\pi^2 v} \left(\frac{k_B}{\hbar}\right)^3 T^3 \int_0^{\theta/T} \frac{\tau_C x^4 e^x}{(e^x - 1)^2} dx \qquad (6.1a)$$

and

$$\kappa_2 = \frac{k_B}{2\pi^2 v} \left(\frac{k_B}{\hbar}\right)^3 T^3 \frac{\left\{\int_0^{\theta/T} (\tau_C/\tau_N) x^4 e^x (e^x - 1)^{-2} dx\right\}^2}{\int_0^{\theta/T} (\tau_C/\tau_N \tau_R) x^4 e^x (e^x - 1)^{-2} dx} \qquad (6.1b)$$

It can be seen that this result is in line with the reasoning that the scattering by N-processes should have an effect, but not the same sort of effect as a clearly resistive process. In the expression for κ_1, N-processes are represented on the same footing as other processes since no distinction is made between them in τ_C. We would, therefore, expect κ_1 to underestimate the conductivity, and there is, indeed, a second term κ_2 which restores some of the 'lost' conductivity.

In interpreting experimental results it is often necessary to use the full expression (eqn (6.1)) (the relative importance of the terms κ_1 and κ_2 for lithium fluoride with different amounts of resistive scattering can be seen in Fig. 8.2), but the numerical analysis is quite lengthy and it is instructive to examine the general results which can be obtained in three limiting cases.

6.1.1. Resistive processes dominant

If a crystal is very imperfect and all modes are strongly scattered by resistive processes, then for all modes $\tau_N \gg \tau_R$ so that $\tau_C \approx \tau_R$. In this case $\kappa_2 \ll \kappa_1$ (this can be seen *qualitatively* by assuming that all the relaxation times are independent of frequency, so that in comparing κ_1 and κ_2 the integrals cancel out and we are left with $\kappa_2/\kappa_1 = \tau_R/\tau_N \ll 1$), and, since $\tau_C \approx \tau_R$, κ_1 is just given by eqn (4.9b) or eqn (4.11a) as though N-processes did not exist. It will be seen later that this relatively simple expression is adequate for analysing the conductivity of crystals which are not too perfect.

6.1.2. Resistive processes present, but N-processes dominant

In this case τ_C is mainly determined by N-processes; $\tau_C \approx \tau_N$ and $\tau_R \gg \tau_N$, and we find that $\kappa_2 \gg \kappa_1$ (again this can be seen qualitatively by assuming frequency-independent relaxation times, from which we obtain $\kappa_2/\kappa_1 = \tau_R/\tau_N \gg 1$). The expression for the thermal conductivity is then

$$\kappa = \frac{k_B}{2\pi^2 v}\left(\frac{k_B}{\hbar}\right)^3 T^3 \frac{\left\{\int_0^{\theta/T} x^4 e^x (e^x - 1)^{-2}\, dx\right\}^2}{\int_0^{\theta/T} \tau_R^{-1} x^4 e^x (e^x - 1)^{-2}\, dx}. \tag{6.2}$$

It is at first surprising that eqn (6.2), which represents the conductivity when N-processes are dominant, does not contain τ_N. However, the effect of N-processes has been taken into account as determining the phonon distribution in the form of eqn (5.2). As N-processes have been assumed to be dominant, the phonon distribution has taken up this displaced form and would not be different for different N-process scattering rates. Thermal resistance then arises from the resistive processes acting on this distribution.

Another interesting aspect of eqn (6.2) appears when we use it to express thermal resistance

$$W = \frac{\int_0^{\theta/T} \tau_R^{-1} x^4 e^x (e^x - 1)^{-2}\, dx}{(k_B/2\pi^2 v)(k_B/\hbar)^3 T^3 \left\{\int_0^{\theta/T} x^4 e^x (e^x - 1)^{-2}\, dx\right\}^2}. \tag{6.3}$$

For a given crystal at some particular temperature the denominator of eqn (6.3) is fixed. Since $\tau_R^{-1} = \sum \tau_i^{-1}$ is the sum of scattering rates for all types of resistive process, we see that $W = \sum W_i$, where the W_i are the thermal resistances which would occur if each of the individual resistive processes i acted separately with N-processes dominant. Thermal resistances are not in general additive, because in κ_1 the relaxation rates add in the denominator of the integral (the combined relaxation *time* is in the numerator) and also, except in the limiting case considered here, κ_2 is too complicated an expression to yield such a simple result.

By using eqn (4.10) to express the functions of x in eqn (6.3) in terms of $C(x)$ and the total heat capacity C, only simple arithmetic is needed to

TAKING ACCOUNT OF NORMAL PROCESSES

show that eqn (6.3) can be written

$$W = \frac{3}{C^2 v^2} \int \tau_R^{-1} C(x)\, dx. \tag{6.4a}$$

This should be compared with eqn (4.11a) which represents the relaxation-time expression for conductivity in the absence of N-processes. In the absence of N-processes the relaxation *time* for each mode is weighted by that mode's contribution to the heat capacity before integrating over all modes to obtain the *conductivity*. If N-processes are dominant, the relaxation *rate* for each mode is weighted by that mode's contribution to the heat capacity before integrating over all modes to obtain the *resistivity*. In the latter case the square of the heat capacity in the denominator of eqn (6.4a) ensures that the resistivity still depends dimensionally on an inverse heat capacity and the conductivity on the first power of a heat capacity.

By writing $v\tau = l$, we can express eqn (6.4a) in terms of mean free paths:

$$W = \frac{3}{C^2 v} \int \frac{C(x)\, dx}{l(x)}. \tag{6.4b}$$

This relation can be compared with eqn (4.11b). We shall see in § 6.2.2 that in the limit of dominant N-processes the variational method leads to exactly the same result as eqns (6.4a) and (6.4b).

There has been one set of experiments on the effect of defects to which this theory of dominant N-processes has been applied directly; this is discussed in § 8.3.1(a).

There is one situation which comes under the present heading to which, however, the Callaway method cannot be applied. If resistive scattering only occurs at the boundaries of a crystal, but N-processes are very frequent, we cannot substitute the value of v/D (D is the appropriate linear dimension of the crystal) for τ_R^{-1} into eqn (6.4a). If we did so we would obtain

$$W = \frac{3}{C^2 v^2} \frac{v}{D} C = \frac{3}{CvD},$$

which is just the resistivity for boundary scattering alone in the absence of N-processes (see § 7.2), and it would appear that the dominant N-processes had no effect. As will be discussed in § 7.3, this particular combination of scattering processes leads to a conductivity *greater* than would result from boundary scattering in the absence of N-processes by an amount directly proportional to the N-process relaxation rate.

6.1.3. Only N-processes acting

While the previous two extreme cases can be approached in practice, we only look at this limit to ensure that the expected result is obtained. We assume that there are no resistive processes, so that $\tau_R \to \infty$ and $\tau_C = \tau_N$.

The denominator of κ_2 then approaches 0 and $\kappa_2 \to \infty$, yielding an infinite conductivity as required.

6.2. The variational method

If there are resistive scattering processes taking place as well as N-processes, then the numerator in the variational expression for the thermal resistivity consists of the sum of the appropriate summations or integrals for each mechanism, two examples of which are given in eqns (4.15) and (5.3). Although in the separate cases to which these numerators were relevant the resistivities could be expressed in an exact form, $1/\kappa = 0$, it is not possible to give a simple expression for the resistivity in the general case of several types of resistive processes acting together, whether or not N-processes are ignored. An outline will be given of the method used by Sheard and Ziman (see Berman, Nettley, Sheard, Spencer, Stevenson, and Ziman 1959) to calculate the conductivity resulting from a combination of point defects and N-processes with neither dominant. It will also be shown that when N-processes are dominant the analysis yields the same result as that obtained in § 6.1.2 by using the Callaway method.

6.2.1. Resistive processes and N-processes both important

In order to illustrate the variational treatment of this general case, the number of terms to be written down will be reduced by taking the resistive scattering to be due only to point defects. This was the assumption used by Sheard and Ziman in applying the method to explain experimental results on phonon scattering by isotopic 'impurities' (see § 8.3.1(a)).

For point defects scattering elastically, combined with N-processes, the variational expression for the thermal resistivity is

$$\frac{1}{\kappa} = \frac{(1/2k_B T^2)[\sum\sum\{\phi(\mathbf{q})-\phi(\mathbf{q}')\}^2 P(\mathbf{q},\mathbf{q}') + \sum\sum\sum\{\phi(\mathbf{q}_1)+\phi(\mathbf{q}_2)-\phi(\mathbf{q}_3)\}^2 P^{\mathbf{q}_1,\mathbf{q}_2}_{\mathbf{q}_3}]}{[1/T \sum \hbar\omega(\mathbf{q})v_G(\mathbf{q})\phi(\mathbf{q})\{\partial \mathcal{N}^0(\mathbf{q})/\partial E(\mathbf{q})\}]^2} ; \quad (6.5)$$

$\phi(\mathbf{q})$ has to be chosen to minimize this resistance.

We have seen that simple forms of $\phi(\mathbf{q})$ can be chosen which make either the first or the second term in the numerator zero. However, if we choose either of these forms of $\phi(\mathbf{q})$, one of the terms in eqn (6.5) will be far from zero. It would be extremely laborious to find the exact form of $\phi(\mathbf{q})$ which would really give the minimum possible value of W, and Sheard and Ziman chose a plausible combination of the two separate forms of $\phi(\mathbf{q})$. The relaxation rate for point-defect scattering increases rapidly with q, and they assumed that above a certain value of q the phonon distribution is governed by the defects and by N-processes, each contributing a term to $\phi(q)$ of the same form as if they were acting alone. For point defects alone the distribution departs very far from equilibrium for small q, and it was assumed that this contribution to $\phi(q)$ reached a limiting value. The trial function was thus different for two ranges of q and was written

$$\phi(\mathbf{q}) = a_0(\mathbf{q}\cdot\mathbf{u}) + \frac{a_4}{q^4}(\mathbf{q}\cdot\mathbf{u}) \quad \text{for} \quad q \geq q_0 = \varepsilon\frac{k_B T}{\hbar v}$$

and

$$\phi(\mathbf{q}) = a_0(\mathbf{q}\cdot\mathbf{u}) + \frac{a_4}{q_0^4}(\mathbf{q}_0\cdot\mathbf{u}) \quad \text{for} \quad q < q_0.$$

It was expected that ε would vary with defect concentration but would be independent of temperature. It was then necessary to vary the coefficients a_0 and a_4 and also ε to find the minimum value of the resistivity given by eqn (6.5).

We would expect the range of q for which the phonon distribution is appreciably influenced by point defects to be widened as the relaxation rate for point-defect scattering increases. Then q_0 and thus ε should decrease. The calculations of Sheard and Ziman showed that for small concentrations of isotopic 'impurities' in lithium fluoride (the theory was initially developed to explain experiments made on this system), the value of ε was ~3, but for mixed crystals with considerable point-defect scattering ε was less than 0·5.

In his original paper on the application of the simple relaxation-time method, Klemens (1951) took N-processes into account by assuming that they prevented the effective relaxation time for point defects from increasing indefinitely for small q by limiting the relaxation time for phonons with $q < k_B T/\hbar v$ to the value of the relaxation time for $q = k_B T/\hbar v = q_0$. The conductivity was given by eqn (4.9) or (4.11) with this modification, so that the integral was split into two parts: for values of q from 0 to q_0 the relaxation time was constant, but from q_0 to q_{max} it varied with q in the normal manner.

Although the Klemens cut-off procedure may not seem to be very different from that of Sheard and Ziman, the numerical results can in many cases be very different. If N-processes are dominant, a wide range of phonons are forced into a distribution such that the first term in the numerator of eqn (6.5) is large. In the limit in which the form of the distribution is determined entirely by N-processes (discussed in the next section) the resistance due to point defects is 55 times greater than is given by Klemens's expression, which reflects the influence of N-processes on the distribution for $q > k_B T/\hbar v$. For a point-defect concentration such that $\varepsilon = 3$, the resulting resistivity is about 20 times greater than the Klemens value. On the other hand, when point defects are much more important and determine the distribution even down to $q = \frac{1}{2}(k_B T/\hbar v)$, there is a wider range of phonons for which the contribution of the first term in eqn (6.5) is negligible, and the resistivity is only a little over half the Klemens value.

6.2.2. Resistive processes present, but N-processes dominant

In this extreme case it is assumed that the phonon distribution is entirely determined by the N-processes and that the defects act on this distribution without influencing it. N-processes thus contribute nothing to the variational expression. For this form of $\phi(\mathbf{q})$ the denominator of eqn (4.15) can be expressed in a simple form. Since

$$\frac{\partial \mathcal{N}^0(\mathbf{q})}{\partial E(\mathbf{q})} = -\frac{\exp(E/k_B T)}{k_B T \{\exp(E/k_B T) - 1\}^2}$$

and

$$\frac{\partial \mathcal{N}^0(\mathbf{q})}{\partial T} = \frac{E \exp(E/k_B T)}{k_B T^2 \{\exp(E/k_B T) - 1\}^2},$$

$$\frac{\partial \mathcal{N}^0}{\partial T} = -\frac{E}{T} \frac{\partial \mathcal{N}^0}{\partial E}.$$

The denominator is

$$\left\{ \sum v(\mathbf{q} \cdot \mathbf{u}) \frac{\partial \mathcal{N}^0(\mathbf{q})}{\partial T} \right\}^2 = \left\{ \frac{1}{3} \frac{\partial}{\partial T} \sum \mathcal{N}^0 vq \right\}^2$$

$$= \left\{ \frac{1}{3} \frac{\partial}{\partial T} \sum \mathcal{N}^0 \frac{\hbar \omega}{\hbar} \right\}^2$$

$$= \frac{C^2}{9\hbar^2}.$$

($|u|$ has been taken as unity since u^2 occurs in numerator and denominator). For isotropic scattering and $P(\mathbf{q}, \mathbf{q}')$ independent of the absolute orientations of q and q', the numerator also comes down to a simple form which can be written (replacing the summation by an integral) as

$$\frac{1}{3\hbar^2 v^2} \int \frac{C(q) \, dq}{\tau(q)}.$$

The thermal resistivity is thus

$$W = \frac{3}{C^2 v^2} \int \tau^{-1}(q) C(q) \, dq$$

which is just the same as the Callaway expression in the same limit of dominant N-processes.

It should be noted that if $\tau(q) \propto q^{-4}$, as for point defects, the expression for the conductivity does not diverge, since the integral is then of the form $\int_0^{\theta/T} \{x^8 e^x/(e^x - 1)^2\} \, dx$ which is finite at both limits. If the scattering is less strongly dependent on q, for example if $\tau(q) \propto q^{-1}$, the simple relaxation-time method does give a finite conductivity because the increase in relaxation time as q decreases does not keep pace with the decrease in the contribution these modes can make to the conductivity, which is governed by the phonon energy and by the density of states (proportional to q^2). The simple relaxation-time expression for κ now contains $\int \{x^3 e^x/(e^x - 1)^2\} \, dx$, while the variational and Callaway expressions for dominant N-processes contain

$$\frac{\left\{\int x^4 e^x (e^x - 1)^{-2} \, dx\right\}^2}{\int x^5 e^x (e^x - 1)^{-2} \, dx}.$$

All these integrals reach limiting values for quite small values of x, and the value of $\kappa_{\text{simple}}/\kappa_{\text{N proc. dom.}}$ is then 1·3. This shows that when resistive scattering still remains important for small q, the addition of dominant N-processes does not increase the resistivity very much.

6.3. The method of Guyer and Krumhansl

Among the treatments which are based on the phonon Boltzmann equation, the work of Guyer and Krumhansl (1966) on phonon hydrodynamics is noteworthy as a relatively early investigation of the general behaviour of phonon systems. The phonon distribution is assumed to depend on both position and time. Taking account of the variation of the distribution across the width of a crystal as well as along its length in a stationary temperature gradient leads to the prediction of Poiseuille flow (see § 7.3), while the introduction of a variation with time leads to the prediction of second sound, which is the propagation of variations in $\mathcal{N}$ according to a wave motion.

The treatment is in terms of an operator form for the Boltzmann equation, and general results are obtained in terms of the collision operators, with a distinction between the operators for normal and resistive scattering, N^* and R^*. The solution of the Boltzmann equation, and hence the expressions for the heat flow and thermal conductivity, are in terms of these operators, and these need only be represented by relaxation rates τ_N^{-1} and τ_R^{-1} to simplify the calculations for numerical applications.

For $N^* > R^*$, corresponding to $\tau_N^{-1} \gg \tau_R^{-1}$, the conductivity is the same as the second term in the Callaway expression and is thus the same as the variational result for the same condition of dominant normal processes. For $R^* > N^*$, corresponding to $\tau_R^{-1} \gg \tau_N^{-1}$, the simple relaxation-time expression (eqn 4.9b) is reproduced, which is also the first term in the Callaway expression.

Guyer and Krumhansl give an expression which covers the whole range of relative values of τ_R^{-1} and τ_N^{-1}:

$$\kappa = \tfrac{1}{3} C v^2 \left(\langle \tau_R \rangle \frac{S}{1+S} + \langle \tau_R^{-1} \rangle^{-1} \frac{1}{1+S} \right) \tag{6.6}$$

where

$$\langle \tau_R \rangle = \frac{\int_0^{\theta/T} \tau_R(x) x^4 e^x (e^x - 1)^{-2} \, dx}{\int_0^{\theta/T} x^4 e^x (e^x - 1)^{-2} \, dx}$$

and $\langle \tau_R^{-1} \rangle$ is a similar average of $\{\tau_R(x)\}^{-1}$. The quantity S is $\langle \tau_N \rangle / \langle \tau_R \rangle$ and is termed the switching factor. If $\tau_N \gg \tau_R$, S is large and $\langle \tau_R \rangle$ is the dominant term in the bracket of eqn (6.6). Resistive processes are dominant, and the conductivity is then just the same as is given by the simple relaxation-time expression. If $\tau_R \gg \tau_N$, S is small and $\langle \tau_R^{-1} \rangle^{-1}$ is the dominant term. Normal processes are dominant in determining the phonon distribution, and the expression for the conductivity is the same as the Ziman or Callaway expression in this limit and resistivities are additive.

Over the intermediate region of relative values of τ_R and τ_N scattering rates deduced from experiments by analyses using the Callaway and Guyer–Krumhansl expressions are very similar, but the fits obtained are a little different where $\langle \tau_R \rangle \approx \langle \tau_R^{-1} \rangle^{-1}$ (Day 1970).

7

THE THERMAL CONDUCTIVITY OF NEARLY PERFECT NON-METALLIC CRYSTALS

SINCE non-metallic crystals have been grown which exhibit all the characteristics to be expected of a perfect crystal, it is quite realistic to divide a discussion of the thermal conductivity into two parts, perfect and imperfect crystals.

An ideal crystal of infinite dimensions would only have finite conductivity because the departures from harmonicity in the lattice vibrations allow interactions among the phonons, and the U-processes which occur introduce thermal resistance. A closer approach to reality comes from assuming that the crystal has finite size and that phonons are scattered resistively from the external boundaries.

At sufficiently high temperatures U-processes are so frequent that they are effectively the only cause of resistance, while at sufficiently low temperatures only boundary scattering need be considered. There is an intermediate region where both U-processes and boundary scattering are important. In a very few crystals there is also a temperature range in which internal resistive processes are negligible but N-processes are dominant, and together with boundary scattering determine the conductivity.

7.1. U-processes

The probability of three-phonon processes occurring contains the population factor discussed in § 5.2, arising from the fact that the rate of change of the number of phonons in mode q_1 depends on the difference between the probability of processes which tend to increase $\mathcal{N}(q_1)$ and those which decrease $\mathcal{N}(q_1)$. This factor can be written as

$$\mathcal{N}_1(\mathcal{N}_3 - \mathcal{N}_2) + \mathcal{N}_3(\mathcal{N}_2 + 1) = \mathcal{N}_1^0(\mathcal{N}_3^0 - \mathcal{N}_2^0) + (\mathcal{N}_1 - \mathcal{N}_1^0)(\mathcal{N}_3^0 - \mathcal{N}_2^0) + \mathcal{N}_3^0(\mathcal{N}_2^0 + 1)$$

where $\mathcal{N}(q_1)$ is abbreviated to $\mathcal{N}_1$, etc., and it has been assumed that for the purpose of calculating the rate of change of $\mathcal{N}$ for a particular mode all other modes can be considered to have their equilibrium population. Substitution of

$$\mathcal{N}^0(\omega) = \frac{1}{\exp(\hbar\omega/k_B T) - 1}$$

and the condition $\omega_1 + \omega_2 = \omega_3$ leads to the factor

$$(\mathcal{N}_1 - \mathcal{N}_1^0)(\mathcal{N}_3^0 - \mathcal{N}_2^0) \tag{7.1}$$

entering the expression for the rate of change in the population of mode q_1.

7.1.1. High temperatures

At sufficiently high temperatures $\hbar\omega/k_B T$ is small for all modes and $\mathcal{N}^0 = k_B T/\hbar\omega$, so that the factor (7.1), with the condition $\omega_1 + \omega_2 = \omega_3$, becomes

$$-(\mathcal{N}_1 - \mathcal{N}_1^0) \frac{k_B T}{\hbar} \frac{\omega_1}{\omega_2 \omega_3}.$$

There are other frequency factors in the expression for the rate of change of $\mathcal{N}$, but for every pair of values of q_2 and q_3 which may make up the three-phonon process with q_1 the contribution to $\partial \mathcal{N}/\partial t$ contains the factor $(\mathcal{N}_1 - \mathcal{N}_1^0)T$. This implies that $\mathcal{N}_1$ returns to $\mathcal{N}_1^0$ with a relaxation time inversely proportional to T. At the high temperatures envisaged, the important phonons have large q and the condition that $\mathbf{q}_1 + \mathbf{q}_2$ must be more than half a reciprocal lattice vector $\mathbf{g}$ for a U-process does not seriously restrict their occurrence. The contribution of a single mode to the heat capacity is the constant k_B when $\omega < k_B T/\hbar$, so that we would expect the high-temperature conductivity to be inversely proportional to T.

It is more difficult to estimate the magnitude of the conductivity from a knowledge of other properties of the crystal. Apart from the mass of an atom (or of a unit cell) and the Debye temperature θ, one factor which enters all simple expressions for heat conductivity is γ^2, the square of the Grüneisen constant. This 'constant' is generally derived as the end product of measurements of thermal expansion, since its behaviour represents a powerful tool for comparing the theory of expansion with experiments. For this purpose, the thermal expansion coefficient β is expressed in terms of the heat capacity per unit volume C_v and the compressibility χ by the relation

$$\beta = \gamma C_v \chi. \tag{7.2}$$

In a derivation by the methods of statistical mechanics of an expression for thermal expansion, γ_i enters as a measure of the rate of change of the frequency of vibration of mode i with the volume of the crystal: $\gamma_i = -\partial \ln \omega_i/\partial \ln V$. The value of γ which then appears in expression (7.2) is a mean value of the γ_i's, weighted towards the γ's for the important modes at any given temperature (many solids have ranges of modes with negative γ's, and at temperatures where these modes are dominant the expansion coefficient is also negative).

We can use again the simple linear chain model of § 4.1.1 to show that all γ's are zero for harmonic force constants. The frequency of a mode (eqn (4.1)) is a function of the force constant, which for harmonic forces is independent of the atomic spacing a and of qa. Since the possible values of q are multiples of $2\pi/Na$, we see that the frequency of the mode which is labelled by the value $q/(2\pi/Na)$ is unchanged as a and the length of the chain are changed. We may take this result over into three dimensions and conclude that γ is zero for each mode if the lattice is harmonic. For a linear chain in which the forces are not harmonic, the relation between ω and q is not as simple as that given by eqn (4.1); the frequency of a mode does depend on the rest position of the atoms and γ is not zero. Therefore γ can be taken as a measure of the departures from harmonicity, and we might expect it to enter an expression for thermal conductivity. In fact, it comes into the perturbation Hamiltonian to the first power and thus appears as γ^2 in the transition probability (negative γ's do not result in negative conductivity!).

For a given phonon there are an enormous number of other modes with which it can interact in a U-process, and the integration of the transition probability over all possible modes is complicated for any model which is at all realistic; not only may the Brillouin zone be of awkward shape, but there are several $\omega - q$ curves to be considered which may show appreciable dispersion. However, various estimates which have been made give the conductivity in the form

$$\kappa \propto M_a a \theta^3 / T \gamma^2 \qquad (7.3)$$

where a^3 is the volume occupied by one atom and M_a is the atomic weight (proportional to the mass of the atom, M). This rather simple dependence on the properties of a crystal has been deduced, with different constants of proportionality, by Akhieser (1939) (see Lawson 1957), Leibfried and Schlömann (1954), Julian (1965), Klemens (1969) and Roufosse and Klemens (1973). Dugdale and MacDonald (1955) proposed a similar expression merely by noting that the dimensionless parameter $\beta \gamma T$ seems to be a measure of the departure of a lattice from harmonicity. They expressed the mean free path as $l = a/(\beta \gamma T)$, a result later derived by Black (1973) for a one-dimensional model. Use of the simple kinetic expression for the conductivity (eqn (3.5)) leads to $\kappa = avC/(3\beta \gamma T)$. From the experimental definition of γ (eqn (7.2)) we obtain $\kappa = av/3\gamma^2 \chi T$. Now $v \sim (\chi \delta)^{-\frac{1}{2}}$ where the density $\delta \sim M/a^3$; also the minimum wavelength of vibrational modes is $\sim 2a$, so that the maximum frequency is $\sim v/2a$, and $k_B \theta = \hbar \omega_{\max} \sim hv/2a$. Making all these substitutions we arrive again at eqn (7.3).

There are thus two aspects of the high-temperature conductivity to study: the temperature dependence and the absolute magnitude. The $1/T$

law should be obeyed if the conductivity is determined by three-phonon processes, and this temperature variation then comes from the variation in the populations of the modes. This origin of the $1/T$ law implies that the volume of the crystal should not be varying, since the nature of the modes will otherwise alter and introduce an extra temperature dependence into the conductivity. However, it is at relatively high temperatures where large volume changes may occur that one looks for a $1/T$ variation of conductivity. These temperatures are often so high that accurate measurements are difficult. The experimental results must therefore be scrutinized rather carefully before too many conclusions are drawn about the validity, or otherwise, of the $1/T$ law for crystals at high temperatures at constant volume. We cannot expect very close agreement between absolute values derived from experiment and from relations as simple as eqn (7.3). Small uncertainties in experimental results are not therefore too unbearable in making this comparison.

7.1.1(a). The $1/T$ law. From its derivation, it can be expected that the characteristic high-temperature behaviour should be observed above the Debye temperature θ. For crystals chosen at random this usually means the region above room temperature, and the $1/T$ law is often seen to be a reasonable approximation to the behaviour at such temperatures. However, out of the wealth of experimental results one can discern that even for the simpler types of crystal, such as the inert gases, alkali halides, silicon, and germanium, the conductivity varies more rapidly than $1/T$. Steigmeier (1969) shows the same trend for a number of III–V semiconducting compounds.

Although part of the observed deviations may be due to the influence of more complicated scattering processes, Slack (1972a) has shown that the change in volume with temperature can produce large departures from the $1/T$ law by itself. In eqn (7.3), a, θ, and also γ vary with volume (even at constant temperature) and this produces a very rapid volume dependence of conductivity, $\kappa \propto v^{-m}$, where according to Slack m is 7 to 8 for some alkali halides and for AgCl and 17 for helium (measured values are similar or even larger). Using such values for the volume dependence, together with the expansion coefficient, Slack found that, if at constant volume $\kappa \propto 1/T$, then the temperature dependence of the conductivity at constant pressure should be $1/T^{1.26}$ for NaCl at 300 K (which is just below its Debye temperature) and $1/T^{2.2}$ for argon (just below θ).

The closest approach to the $1/T$ law should then occur for measurements made at constant volume, and there are such measurements on solid argon by Clayton and Batchelder (1973). If argon is crystallized inside a metal tube which has small thermal expansion, the pressure of formation can be chosen so that even when the solid has cooled to the lowest temperature at which measurements are made it will not shrink away from the sides of the tube. Any density change then results only from the contraction of the containing tube, and from the effects on it of the pressure variations caused by the argon changing in temperature.

Clayton and Batchelder found that at constant volume the conductivity between 20 and 150 K ($\theta/4$ to 2θ) obeys the $1/T$ law quite well as shown in Fig. 7.1. Julian (1965) has indeed predicted that this law should be obeyed down to $\theta/4$ for the inert gas crystals with face-centred cubic lattices.

Clayton and Batchelder show a comparison of their results with those of Krupskii and Manzhely (Manzhelii) (1967, 1968) which were made at effectively zero pressure and thus for volumes which were allowed to vary according to the expansion coefficient of the argon itself. At 25 K the density of the Krupskii and Manzhely crystals was close to that of one of the Clayton and Batchelder specimens and the conductivities were similar in magnitude. The conductivity of the free-standing crystal varied with temperature as $T^{-1.3}$. At 75 K, however, the density for a crystal at zero pressure has become 7 per cent smaller, and, as predicted by the straightforward corrections to eqn (7.3), the conductivity is less than $\frac{2}{3}$ of the value for a crystal forced to retain the 25 K density. The temperature variation at 75 K is $T^{-2.0}$, as predicted by Slack from the effects of density changes alone. It thus seems unnecessary to invoke higher-order phonon scattering processes to explain even this great departure from the $1/T$ law.

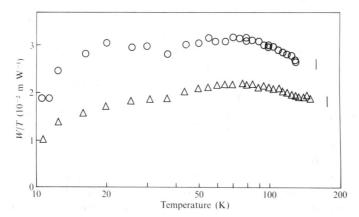

FIG. 7.1. Temperature dependence of WT^{-1}, the thermal resistivity divided by temperature, for specimens of solid argon grown at two different densities. The pressure cell changes its volume slightly with temperature and the estimated corrections to the conductivity for this volume effect would make WT^{-1} at constant volume even more nearly constant. However, the corrections are not sufficiently certain and the improvement is so small that the actual experimental values are shown here. (After Clayton and Batchelder 1973.)

There are some crystals for which an appreciable departure from the $1/T$ law cannot be explained so simply. For example, the conductivities of germanium and silicon at high temperatures vary approximately as $T^{-1\cdot 3}$. However, only a small correction (0·02) in the power of the temperature dependence can, according to Slack, be ascribed to the effects of thermal expansion. It is possible that extra scattering over the whole phonon spectrum due to four-phonon processes, which would provide an additional thermal resistivity proportional to T^2, may be responsible for this temperature variation (see § 13.1). However, Klemens and Ecsedy (1976) have proposed an explanation which involves four-phonon processes in a slightly less straightforward manner. As was pointed out by Pomeranchuk (1941), there are severe restrictions on the kinds of three-phonon processes in which low-frequency longitudinal phonons can participate. The contribution of these phonons alone is then limited by four-phonon processes. The total conductivity can be considered as the sum of a contribution from the low-frequency longitudinal phonons, which Klemens and Ecsedy show to be proportional to $T^{-\frac{3}{2}}$, and a contribution from all other phonons proportional to T^{-1}. Even at constant volume the measured conductivity would, therefore, vary faster than T^{-1}.

Ranninger (1965) has suggested that even at constant volume there will be a decrease in phonon frequencies with increasing temperature, so that there could be quite considerable departures from the $1/T$ law although only three-phonon processes may be important.

7.1.1(b). Absolute magnitudes. There have been a number of comparisons between predicted and measured absolute values of the thermal conductivity of pure non-metallic crystals in the temperature region where the $1/T$ law should hold. The present situation can be judged from the work of Slack (1977), which is mainly concerned with face-centred cubic lattices. He started from eqn (7.3) with the value of the constant of proportionality as given by Julian (1965). This constant itself is weakly dependent on the value of γ, and varies by ~10 per cent of its value for $\gamma = 2$ as γ varies from 1 to 3. Slack used the value of the constant for $\gamma = 2$ (a reasonable average value for many common crystals) which is then $3\cdot 0 \times 10^{-5}$ in MKS units, with a expressed in nm ($3\cdot 0 \times 10^{-8}$ if conductivity is to be in W cm^{-1} K^{-1} and a in Ångstrom units). He then considered the values of θ, γ, and M_a to be taken when there is more than one kind of atom and more than one atom per primitive unit cell. It can generally be assumed that the contribution of the optic modes is negligible, because even if their energy is low enough for them to be excited their group velocity is usually low. We therefore require the Debye temperature and Grüneisen constant appropriate to the acoustic modes.

For the acoustic modes the conductivity is regarded as the sum of contributions from each sub-lattice, where each set suffers scattering by phonons appropriate to all other sub-lattices. If $\bar{M}_a$ is the mean atomic weight of all the constituent atoms, this consideration introduces a factor $\nu^{\frac{1}{3}}$ into the conductivity when there are ν atoms per unit primitive cell and a^3 is now the average volume occupied by one atom. Slack showed how to calculate $\tilde{\theta}$, the value of the Debye temperature appropriate to the acoustic phonons, and the value of $\tilde{\theta}_0$ appropriate to $T = 0$ K is given by $\tilde{\theta}_0 = \nu^{-\frac{1}{3}}\theta_0$. The value of $\tilde{\theta}_\infty$ appropriate to high temperatures is generally similar to $\tilde{\theta}_0$ or slightly lower. Slack computed values of $\tilde{\theta}_\infty/\tilde{\theta}_0$ for crystals with well-enough known phonon spectra and used these ratios to deduce $\tilde{\theta}_\infty/\tilde{\theta}_0$ for other crystals, for which the necessary information was not available.

The calculation of $\tilde{\gamma}$ appropriate to acoustic phonons is even more difficult. For crystals with the diamond structure, other than diamond itself, a single value $\tilde{\gamma}_\infty = 0.58$ corresponding to the mean of values calculated for germanium and zinc telluride was taken, while for diamond $\tilde{\gamma}_\infty$ was taken as 0.77 times the high-temperature value of γ deduced from the measured thermal expansion. For all other crystals considered, the high-temperature thermal expansion value of γ was used, although it is known to differ appreciably from the required $\tilde{\gamma}_\infty$.

Slack compared calculated and observed conductivities at the high-temperature value of θ appropriate to acoustic phonons. Putting $T = \tilde{\theta}_\infty$, the expression for the conductivity at this temperature becomes

$$\kappa(\tilde{\theta}_\infty) = 3 \cdot 0 \times 10^{-5} \nu^{\frac{1}{3}} \bar{M}_a a (\tilde{\theta}_\infty)^2 (\tilde{\gamma}_\infty)^{-2} \text{ W m}^{-1} \text{ K}^{-1}.$$

Figs 7.2, 7.3, and 7.4 show this comparison for crystals with $\nu = 1, 2$, and more than 2, respectively. On the whole the agreement is very good, even when ν is very large.

Slack pointed out that his expression for the conductivity can be approximated by

$$\kappa = \frac{3 \cdot 0 \times 10^{-5} \bar{M}_a a \theta_0^3}{T \gamma^2 \nu^{\frac{2}{3}}} \text{ W m}^{-1} \text{ K}^{-1} \qquad (7.3a)$$

and the agreement between calculated and experimental values is not seriously different if we use this relation, taking θ_0 from low-temperature heat capacity measurements and γ from the high-temperature limiting thermal expansion values in all cases.

Roufosse and Klemens (1973) also considered the effect of the number of atoms in the unit cell on the conductivity. They concluded that the relatively simple expression proposed by Slack, which did not take into account the details of the U-process scattering but gives quite good agreement with experiment, resulted from the partial cancellation of two opposite effects. With an increase in the number of atoms per unit cell,

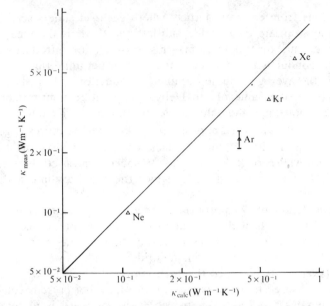

FIG. 7.2. Comparison of the calculated and measured conductivities of cubic crystals at $\bar{\theta}_\infty$, the high-temperature value of the Debye temperature corresponding to acoustic phonons, for one atom per unit cell. (After Slack 1977.)

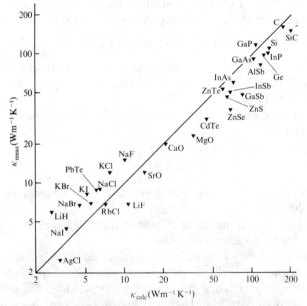

FIG. 7.3. Comparison of the calculated and measured conductivities of cubic crystals at θ_∞, the high-temperature value of the Debye temperature corresponding to acoustic phonons, for two atoms per unit cell. (After Slack 1977.)

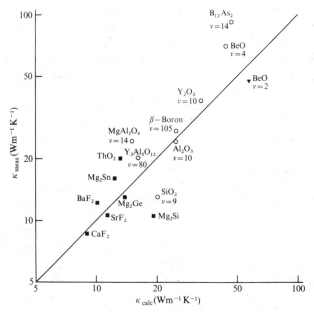

FIG. 7.4. Comparison of the calculated and measured conductivities of cubic crystals at $\tilde{\theta}_\infty$, the high-temperature value of the Debye temperature corresponding to acoustic phonons, for more than two atoms per unit cell. (After Slack 1977.)
■ $\nu = 3$; ○ $\nu \geq 4$

the interaction matrix element for U-processes is reduced, but it is necessary to sum over a greater number of relevant reciprocal lattice vectors.

It can be seen that the agreement between calculated and measured conductivities for the inert gases with $\nu = 1$ is neither better nor worse than for many crystals with $\nu = 2$. The sensitivity of the conductivity of argon to volume has already been discussed, and one must be careful that values of the appropriate properties at the same volume are always used. For example, the appropriate Debye temperature for argon, $\tilde{\theta}_\infty$, derived from measurements on a free-standing crystal near 0 K would be 84 K, but by this temperature the density has decreased by about 6 per cent, so that $\tilde{\theta}_\infty$ would be quite different.

Such considerations are bound to limit the agreement between predicted and measured conductivities, even when it is not affected by more subtle effects due to particular features in the phonon spectra.

Blackman (1935) showed that in crystals containing more than one atom per unit cell the conductivity should be affected by the mass ratio of the constituent atoms, and he ascribed the observed trend in the heat conductivity of the alkali halides to this cause. The mechanism of this effect has also been discussed by Ziman (1960).

If any contribution to the conductivity from the optic phonons can be neglected, then their influence on the conductivity arises from their scattering of acoustic phonons. On a simplified model of acoustic modes with no dispersion and of optic modes all of the same frequency, there are no U-processes involving optic modes for nearly equal masses ($\omega_{\max ac} \sim \omega_{opt}$). As the mass ratio increases and ω_{opt} becomes greater than $\omega_{\max ac}$, the number of possible U-processes increases until the mass ratio is so great that $\omega_{opt} \gg \omega_{\max ac}$; energy conservation between two acoustic mode phonons and an optic phonon cannot be satisfied and the number of U-processes decreases again. The ratio $\kappa_{exp}/\kappa_{calc}$ would therefore be expected to decrease between mass ratios of 1 and ~ 4 (when on a simple model $\omega_{opt} = 2\omega_{\max ac}$) and then to increase, where κ_{calc} is the conductivity derived from an expression such as eqn (7.3) which neglects any influence of the optic modes.

This trend is fairly well illustrated by the alkali halides, and Fig. 7.5 shows the ratio $\kappa_{exp}/\kappa_{calc}$ at the Debye temperature corresponding to the acoustic modes, taken mainly from the work of Slack (1977). Recent measurements by Moore, Williams, and Graves (1975) replace earlier data of Eucken and Kuhn (1928) on RbBr and RbI and confirm the values for RbCl quoted by Slack. In view of the disagreement between Eucken and Kuhn's results and more modern measurements on some

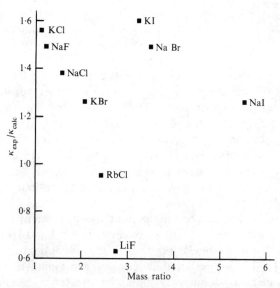

FIG. 7.5. The ratio of the measured and calculated conductivities of some alkali halides at $\bar{\theta}_\infty$, the high-temperature value of the Debye temperature corresponding to acoustic phonons, as a function of the ratio of the masses of the two constituent atoms. (Data from Slack 1977.) $\kappa_{exp}/\kappa_{calc} \sim 1 \cdot 9$ for RbBr and RbI (mass ratios 1·07 and 1·48).

alkali halides, KF has not been included in Fig. 7.5 (even though it seems to fit quite well) as the only measurements are by Eucken and Kuhn. Since the measured conductivities vary faster than $1/T$, little weight can be put on the exact values of the ratios $\kappa_{exp}/\kappa_{calc}$, but the trend would not be materially altered by choosing some other characteristic temperature at which to make the comparisons. Slack points out that for some alkali halides there are optic phonons with group velocities as great as those of acoustic phonons, so that there may be a direct contribution from them to the heat conductivity. However, experimental evidence for this is conflicting.

Other series of crystals in which the influence of the mass ratio might be sought are the III–V and II–VI semiconductors. Using Slack's tabulated values again for the former and his measurements and tabulations for the latter (Slack 1972a), even less trend in the ratios $\kappa_{exp}/\kappa_{calc}$ can be found (see § 13.1).

The extent to which an expression such as eqn (7.3), modified in accordance with the mass ratio, can in general predict the conductivity of a crystal is thus not really clear. For the alkali halides, use of an average value of ~ 1.4 for $\kappa_{exp}/\kappa_{calc}$ would yield conductivities within ~ 50 per cent of those observed, an uncertainty which would be reduced if the mass effect were taken into account. For the III–V compounds, an average value of ~ 0.8 for the ratio would yield conductivities somewhat closer to those observed and this would not be improved by invoking a mass effect.

Slack (1973) has used eqn (7.3) to discuss which non-metallic crystals would be expected to have very high thermal conductivities at room temperature. As θ enters to the third power it is important to have a high value for this, even though high θ usually implies small $\bar{M}_a$ (for high θ there should be strong bonding between light atoms, in fact $\bar{M}_a \theta^2$ correlates quite well with hardness as measured on the Mohs scale). We also require ν and γ to be small. These combined criteria are best met by adamantine crystals, which have the diamond structure, and Slack lists twelve of these which should have conductivities of at least $100 \text{ W m}^{-1} \text{K}^{-1}$ at 300 K (the thermal conductivity of copper at 300 K is $\sim 400 \text{ W m}^{-1} \text{K}^{-1}$). Measurements have been made on pure single crystals of six of these materials and on poorer specimens of two. The comparison between predictions and experimental values is shown in Fig. 7.6, where the variations in γ have been omitted from the scaling parameter. The estimated conductivities for the crystals which have not been measured are as follows ($\text{W m}^{-1} \text{K}^{-1}$ at 300 K): BeS, 300; BAs, 210; GaN, 170; AlP, 130. As an example of the benefit of the small values of γ for adamantine compounds, Slack quotes LiH which has a relatively small conductivity at 300 K although $\bar{M}\theta^2$ is high. However, γ^2 is about four times greater than for most of the adamantine crystals considered.

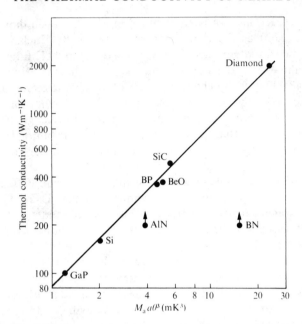

FIG. 7.6. The thermal conductivity at 300 K of various adamantine crystals as a function of the scaling parameter. The specimens of AlN and BN were of poor quality, so that the intrinsic conductivities are expected to be higher than those measured. (After Slack 1973.)

7.1.2. Low temperatures

7.1.2(a). U-processes and the exponential temperature variation. The $1/T$ law was derived from expression (7.1) by assuming that for all the important modes $\mathcal{N}^0 \propto T$, and that these modes were of such large q that in a three-phonon interaction U-processes were just as likely as N-processes. In either Fig. 5.1b or Fig. 5.1c, if the phonon concerned has its wave vector reaching nearly to the zone boundary, interactions with another phonon are about equally likely to produce a resultant which lies inside or outside the Brillouin zone, since the main determining factor is the direction of the second phonon.

At low temperatures heat is mainly transported by phonons of low q, because phonons with $\hbar\omega \sim 4k_B T$ provide the greatest contribution to the heat capacity (the maximum value of $x^4 e^x/(e^x - 1)^2$ occurs for $x = 3\cdot 8$). For such phonons $\mathcal{N}^0$ cannot be approximated as proportional to T. In addition, out of all the interactions which can occur with other phonons, only rarely will the second phonon have a large enough q for the resultant to fall outside the zone. The proportion of phonon–phonon interactions which are U-processes thus decreases at low temperatures.

If $\mathbf{q}_1$ is a small wave vector reaching only a small fraction of the way to the zone boundary, it is necessary for $\mathbf{q}_2$ to be a large vector for a U-process. The smallest value of q_2 will depend on the direction of q_1 in relation to the shape of the zone, but in any case its magnitude will be comparable with a reciprocal lattice vector. This means that ω_2 will be about the maximum that can occur in the crystal. For a linear dispersion curve ($\omega \propto q$) this frequency would be $\sim k_B \theta/\hbar$, but for real dispersion relations the maximum usually differs appreciably from this value. If the temperature is sufficiently small,

$$\mathcal{N}^0(\omega_2) = \frac{1}{\exp(\hbar\omega_2/k_B T) - 1}$$
$$\sim \exp(-\hbar\omega_2/k_B T).$$

As ω_1 is assumed to be small, $\omega_3 \sim \omega_2$, and expression (7.1) becomes

$$(\mathcal{N}_1 - \mathcal{N}_1^0)\frac{\partial \mathcal{N}^0(\omega_2)}{\partial \omega_2}\omega_1 = -(\mathcal{N}_1 - \mathcal{N}_1^0)\frac{\hbar\omega_1}{k_B T}\exp\left(-\frac{\hbar\omega_2}{k_B T}\right).$$

Since $\hbar\omega_2/k_B$ is of the order of θ, we see that the relaxation time for phonons q_1 will contain a factor $\exp(+\theta/bT)$, where b is of the order of unity. The complete expression for the relaxation time at low temperatures cannot be derived in this simple-minded way, and contains additional powers of frequency and temperature as well as the characteristic exponential term. Because of the strong temperature dependence of the exponential at low temperatures, the powers of T which should precede it in an expression for the conductivity cannot be determined unequivocally by experiment. The power usually derived from theory is small (see, for example, Leibfried and Schlömann 1954, Klemens 1958), but Julian (1965) obtains the form $(T/\theta)^8 \exp(\theta/bT)$ for the inert gas crystals. The form

$$\kappa \propto T^\xi \exp(\theta/bT) \quad (7.4)$$

with ξ and b both of the order of unity was given originally by Peierls (1929) long before there was any experimental evidence for a dramatic change from the pedestrian $1/T$ law.

Although the exponential relation was looked for soon after Peierls's paper appeared, it was about 20 years before it was clearly found. There were two reasons for this, apart from the intervention of the war: the appropriate temperature range for its clear manifestation is restricted to roughly $\frac{1}{30} < T/\theta < \frac{1}{10}$, so that the crystal and the temperature range of the measuring apparatus have to be suitably matched, and the crystal must be very perfect so that U-processes really can dominate the conductivity. For this purpose, perfection includes absence of isotopes of the constituents of the crystals (see §8.3.1(a)). By chance the early measurements were on SiO_2, KCl, and other crystals with elements consisting of several isotopes, while again by chance some of the post-war measurements were on materials with very low concentrations of isotopes in nature, such as ^{4}He and Al_2O_3 (see Berman, Foster, and Ziman 1956).

Both Leibfried and Schlömann, and Klemens derived expressions for the low-temperature conductivity which contain the factor $(M_a a T^3)/(\gamma^2 \theta)$ before the exponential. The high-temperature conductivity is proportional to $M_a a \theta^3/\gamma^2 T$, so that if κ_θ is the conductivity at $T = \theta$, κ at any other 'high' temperature is $\kappa_\theta \theta/T$ and the low-temperature conductivity is

$$\kappa \propto \kappa_\theta \left(\frac{T}{\theta}\right)^3 \exp\left(\frac{\theta}{bT}\right)$$

(ignoring any distinction between θ's appropriate to high and low temperatures). For most real crystals which have been measured at sufficiently low temperatures for such a law to be expected to hold, there is no temperature range in which other scattering mechanisms do not affect the conductivity. The resistance due to U-processes cannot then be deduced merely by looking at the experimental results, and can only be inferred from detailed analysis. Sufficiently perfect crystals of a few materials have been measured in which U-processes dominate the conductivity over a relatively wide temperature range. By plotting $\log(\kappa T^{-\xi})$ against $1/T$ for various values of ξ we might hope to find which value of ξ leads to the best straight line.

Fig. 7.7 shows plots of $\log(\kappa T^{-\xi})$ against $1/T$ for different values of ξ for NaF, using the data of Jackson and Walker (1971). The values of ξ are 0, 1, 2, and 3, and it can be seen that the quality of the fit is not noticeably different for any of these values of ξ. A similar set of curves can be plotted for LiF using the measurements of Thacher (1965), and again it is impossible to state which value of ξ best represents the data. For each value of ξ an appropriate value of b can be derived from the slope of the corresponding curve. The combinations of ξ and b which fit the results for NaF and LiF are shown in Table 7.1. These are both face-centred cubic crystals; the situation in hexagonal close-packed helium crystals will be discussed in the next section.

7.1.2(b). Anisotropy of U-process-dominated conductivity. Many crystals do not have cubic lattices and their symmetry is reflected in the shape of the Brillouin zone. As an example, Fig. 5.1c shows a cross-section of a zone which is rectangular. A larger resultant q is required for a U-process if the directions of q_1 and q_2 are predominantly in the q_x direction than if they are in the q_y direction.

For heat flow along one of the principal axes, the energy is mainly carried by phonons directed along that axis. The smallest-energy high-frequency phonons with which these thermal phonons can interact in a

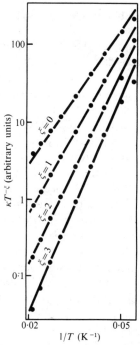

FIG. 7.7. Plots of thermal conductivity multiplied by various powers of the temperature as a function of inverse temperature for NaF. The absolute values of the ordinate change by 10 between each curve. (Data from Jackson and Walker 1971.)

U-process are then also directed along the same axis. We would thus expect the exponentials determining the conductivities in the two directions to have exponents in the ratio of the two maximum frequencies corresponding to the maximum q_x and q_y. If the phonon velocities were the same in the two directions and there were no dispersion, this would

TABLE 7.1

Different forms of the temperature dependence of U-process-limited thermal conductivity which fit experimental results equally well†

NaF	LiF
	$T^{-1} \exp(\theta/6 \cdot 0T)$
$T^0 \exp(\theta/3 \cdot 5T)$	$T^0 \exp(\theta/4 \cdot 1T)$
$T^1 \exp(\theta/2 \cdot 9T)$	$T^1 \exp(\theta/3 \cdot 3T)$
$T^2 \exp(\theta/2 \cdot 4T)$	$T^2 \exp(\theta/2 \cdot 6T)$
$T^3 \exp(\theta/2 \cdot 2T)$	$T^3 \exp(\theta/2 \cdot 4T)$

† The forms for NaF are derived from Fig. 7.7; those for LiF are derived from similar plots.

also be the ratio of the two maximum q's. Such an effect was predicted by Simons (1960).

In real crystals the velocities and the degrees of the dispersion are different in different directions, so that such a simple anisotropy would not be expected. Nevertheless, since the exponents in the exponentials are anisotropic, then regardless of the niceties of the calculation we would expect the resulting anisotropy to be very dependent on temperature.

The hexagonal close-packed form of helium is the only non-cubic crystal which has been measured in a sufficiently perfect form for U-processes to dominate the conductivity completely at relatively very low temperatures. The different exponents can then produce a large difference between the exponentials. Mezhov-Deglin (1964) suggested anisotropy as a possible reason for the different conductivities found for different crystals grown under apparently identical conditions. The correlation between conductivity and direction was inferred by Hogan, Guyer, and Fairbank (1969) from measurements on a large number of single crystals, and this was confirmed later by simultaneous measurements of orientation and conductivity by Berman, Day, Goulder, and Vos (1973). The extreme conductivities for a crystal with molar volume $18 \cdot 6 \times 10^{-6} \, \text{m}^3 \, (\text{mole})^{-1}$, grown at a pressure of 85 atm, are shown in Fig. 7.8 (the conductivity in a direction at 7° to the c-axis is very nearly the same as along the c-axis). The conductivities would be equal at ~ 3 K and are equal again at $0 \cdot 7$ K (where boundary scattering dominates, see § 7.2), but between these two temperatures the anisotropy has a maximum of $\sim 20:1$ at ~ 1 K, the conductivity in directions perpendicular to the crystalline c-axis being the larger within this temperature range.

We meet again the difficulty of deciding the best values of ξ in eqn (7.4) to represent the measured conductivities. In this case, not only does the choice of ξ affect the value deduced for b, but it also affects the derived ratio of $b_\parallel$ to $b_\perp$. If both ξ's are taken as zero (quite a satisfactory fit) the ratio of the b's is about 1·8, but is 2 or more if different values of ξ are chosen for the two directions, judged solely by which values appear to give the best fits to the experimental points obtained for the two directions. It is clear that any plausible fit will give $b_\parallel/b_\perp$ greater than the ratio 1·4 of the limiting phonon frequencies derived from the measured dispersion curves for the lowest-lying phonon modes in the two directions.

Benin (1973) has considered the helium problem in more detail than is appropriate here. He showed that interactions involving a low-lying transverse optic mode, which at large q lies below the transverse acoustic mode in the direction perpendicular to the c-axis, has an important effect on the conductivity $\kappa_\perp$. By making a simplified model of the phonon dispersion curves, Benin was able to explain the main features of the anisotropy.

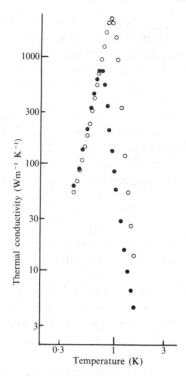

FIG. 7.8. Anisotropy of thermal conductivity of helium associated with anisotropy in the probability of U-processes: ○ heat flow in the plane perpendicular to the c-axis; ● heat flow in a direction making an angle of 7° with the c-axis. (Data from Hogan *et al.* 1969.)

7.2. Boundary scattering

The occurrence of both T and θ in so many expressions for the thermal conductivity, often in the form of the ratio T/θ, produces conductivity curves for different crystals which are similar when plotted as κ versus T/θ. For example, the exponential behaviour is observed below $\sim \theta/10$, which can vary from ~ 200 K for diamond down to ~ 2 K for helium grown at low pressure. For crystals which have dimensions of the order of millimetres, the exponential rise in conductivity comes to a fairly abrupt end at $\sim \theta/30$, and below this temperature it decreases again rather rapidly. This fall in conductivity is due to scattering of phonons at the external boundaries of the crystal. Peierls (1929) had predicted that such a process would prevent the conductivity from increasing indefinitely at low temperatures, and the effect was first observed by de Haas and Biermasz (1935).

Casimir (1938) calculated the heat flow to be expected if the only interactions undergone by phonons were diffuse scattering at the boundaries. He treated the flow of phonons under these conditions as analogous to the flow of radiation down a tube having diffusely scattering walls (he pointed out that in the experiments of de Haas and Biermasz the important phonon wavelengths, even at the lowest temperatures, would be small compared with the roughness of the surface). He showed that the heat flow is proportional to the cube of the radius, to the temperature gradient, and to T^3. At the temperatures concerned, the specific heat of a crystal is also proportional to T^3, and the heat flow can be expressed in terms of the heat capacity per unit volume by a simple expression, if we use an appropriate average phonon velocity:

$$H = \tfrac{2}{3} C \bar{v} \pi r^3 \frac{dT}{dz} \tag{7.5}$$

where r is the crystal radius, and the average velocity $\bar{v}$ is equal to $v_1(2s^2+1)/(2s^3+1)$, where s is the ratio of longitudinal to transverse phonon velocities v_1/v_2.

The definition of thermal conductivity derived from longitudinal heat flow along a rod, as given by eqn (2.3), relies on the heat flow being proportional to the cross-sectional area of the rod. This will be true if the resistance is caused throughout the body of the crystal, but we are now assuming just the opposite condition in which there is no scattering within the crystal and the flow is entirely determined by what happens at the boundaries. If we wish to derive an equivalent conductivity by applying eqn (2.3) to a heat flow which obeys eqn (7.5), we find that for boundary scattering this 'conductivity' is proportional to the radius of the crystal, and, if slight differences between different averages of phonon velocities are neglected, corresponds to an effective mean free path $2r$, which is the diameter of the crystal.

We could have arrived at the same result by assuming that there is a meaningful conductivity still given by the simple kinetic expression (eqn (3.5)), where the mean free path has become limited to the diameter of the crystal $D = 2r$. Since $C \propto T^3$, we again find that the 'conductivity' is proportional to T^3 and to the diameter of the crystal. Although this argument gives the same result for the heat flow as Casimir's calculation, it is not obvious that we can apply the kinetic expression to a situation for which a true conductivity cannot be defined, nor that the mean free path should be exactly the diameter of the crystal.

It is instructive both for boundary scattering and for Poiseuille flow of phonons (§ 7.3) to draw an analogy with gas flow through a tube. For flow at very low pressures we can assume that gas molecules never collide with one another, and the flow is only determined by the way in which they are scattered at the walls of the tube through which they are flowing.

Knudsen (1909) first calculated the flow under these conditions, which bear his name. We can express the amount of gas flowing through a long tube per second, G, by the value of pV, the product of its pressure and volume:

$$G^{pV}_{\text{Knud}} = \tfrac{2}{3}\bar{v}\pi r^3 \frac{p_1 - p_2}{L} \qquad (7.6)$$

where $\bar{v}$ is the average thermal velocity of the molecules and $p_1 - p_2$ is the pressure drop over a length L of the tube. The flow is again proportional to the cube of the tube radius and the expression is very similar to eqn (7.5). Indeed, at these low pressures the pressure gradient is constant, so that we could replace $(p_1 - p_2)/L$ by dp/dz to make the analogy even closer.

For gases the opposite familiar flow regime leads to Poiseuille flow. At normal pressures the flow is determined by intermolecular collisions, and for streamline flow with zero drift velocity at the walls the flow rate is given by the Poiseuille expression

$$G^{pV}_{\text{Pois}} = \frac{\pi r^4}{8\eta}\bar{p}\frac{p_1 - p_2}{L} \qquad (7.7)$$

where $\bar{p}$ is the mean pressure $(p_1 + p_2)/2$ and the viscosity η is a measure of the frequency of intermolecular collisions and is thus related to the molecular mean free path. If measurements of gas flow were made at very low pressures, where there are no intermolecular collisions, it would not be sensible to try to derive a corresponding viscosity, which is a measure of the frequency of such collisions. If one did so, the recipe derived from eqn (7.7) would be to divide $\tfrac{1}{8}\pi r^4 \bar{p}(p_1 - p_2)/L$ by the measured flow. As the flow actually obeys eqn (7.6), one must arrive at the result that the viscosity is proportional to the radius of the tube. It has, however, become customary in the case of heat conduction to derive values for the 'conductivity' even when the flow is not proportional to the cross-sectional area, by dividing the heat flow by this area. Such a conductivity is proportional to the radius of the specimen measured and is not an intrinsic property of the material.

Boundary scattering has been observed in an enormous number of crystals, and a very clear demonstration of the dependence of flow rate on the cube of the linear dimension of the cross-section is shown in Fig. 7.9, taken from the work of Thacher (1965). If the heat flow is proportional to T^3 and to r^3, then κ/T^3 should be proportional to r (Thacher's specimens had nearly square cross-sections and r is replaced by the mean side of the rectangular cross-sections).

Boundary scattering severely restricts the thermal conductivity of materials consisting of compacted microcrystalline aggregates. In many cases

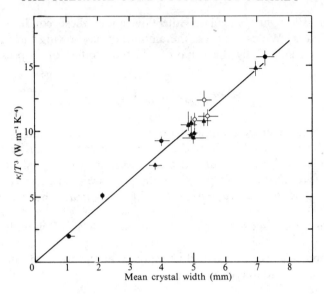

FIG. 7.9. The variation of conductivity with crystal size for LiF, plotted as κ/T^3 versus mean side of cross-section. (After Thacher 1965.)

the conductivity can be described by a phonon mean free path of the same order of magnitude as the crystallite size. If, for example, the crystallites are of micron size, the conductivity of the material at low temperature will be about one-thousandth of the 'conductivity' of a single crystal of millimetre dimensions. Even in amorphous materials, such as glasses (see Chapter 9), phonon mean free paths become appreciable at low temperatures, so that their conductivity can exceed that of polycrystalline materials which may have quite high conductivities at normal temperatures. An example of such a comparison between the conductivities of polycrystalline materials and amorphous substances is shown in Fig. 7.10. This simple description only applies if the polycrystalline material has nearly single-crystal density. Otherwise the contact resistance between the particles introduces a further overall resistivity.

Casimir's calculation was made for an infinitely long crystal with perfectly rough walls. Real crystals are not only finite in extent but often have surfaces which cannot be described as perfectly rough, so that some specular reflections of phonons may occur. For gas flow von Smoluchowski (1910) showed that, if one can consider that a fraction F of the collisions with the walls are diffuse and a fraction $(1-F)$ are specular, the flow in a long tube is increased by a factor $(2-F)/F$. The same result is obtained for the increase in heat flow in a long crystal, but F, and consequently the factor by which the flow is increased, depends on

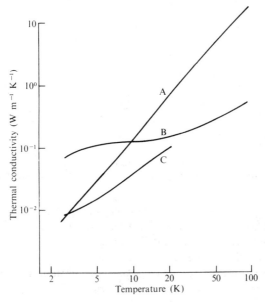

FIG. 7.10. The thermal conductivities of polycrystalline graphite (30 nm crystallites) (curve A), vitreous silica (curve B), and nylon (curve C). (After Berman 1953.)

temperature. Whether a surface appears rough or smooth to phonons depends on the ratio of the size of the asperities to the wavelength of the phonons interacting with the surface. The dominant phonon wavelengths increase with decreasing temperature, so that as the temperature decreases the surface appears to become smoother.

The effect of a crystal having finite length can be understood if we consider the heat flows in a long and a short crystal of the same cross-sectional areas and with the same temperature gradients over the central portions where ΔT and L are measured. We assume that for the long crystal a temperature gradient extends almost indefinitely beyond the region of the thermometers, so that some phonons coming from the high-temperature end can have arrived, without collisions, at the measuring region from regions at much higher temperature. For the short crystal they can only have come from the warm end of the crystal which is scarcely hotter than the measuring region, being close to it. In the radiation analogue we may represent the flow in the short black tube as coming from a long tube whose temperature remains constant beyond the position of the real ends. In a long tube the radiation comes from regions that are increasingly warmer as we go further from the measuring region. The rough form of the temperature gradients in the two cases is shown in Fig. 7.11 (for a finite tube the temperature gradient cannot be exactly linear right up to the ends).

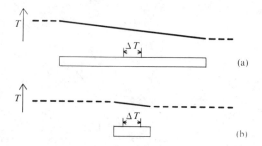

FIG. 7.11. The effective temperature distributions along (a) long and (b) short specimens which have the same temperature gradients over the central region.

The length effect and the influence of specular reflection and its temperature variation were first described by Berman, Simon, and Ziman (1953) and Berman, Foster, and Ziman (1955). Fig. 7.12 shows a comparison between the low-temperature conductivities of crystals of Al_2O_3, which differ only in that one of each pair was ground and the other flame polished (the crystals were long enough for the length effect to be small, so that the ratio of the conductivities represents the factor $(2-F)/F$ and its temperature variation).

Except at the lowest temperatures, boundary scattering occurs together with other resistive processes which scatter throughout the crystal. The relaxation rates for the latter processes are generally dependent on frequency, but the 'thermalizing' effect of rough surfaces prevents phonons which are weakly scattered inside the crystal from running away with the heat flow. The way of combining a resistive process which occurs only at the boundaries with others occurring throughout the crystal was discussed by Herring (1954) and later by Ziman (1960), Hamilton and Parrott (1969), and Srivastava and Verma (1973).

It appears that for a perfectly rough crystal little error is, in fact, involved in adding a relaxation rate v/D for boundary scattering to the resistive relaxation rates for internal processes. For a smooth crystal the equivalent relaxation rate is dependent on frequency. In both cases, because of the nature of the conductivity integral, the combined heat resistivity cannot be regarded as the sum of separate defect and boundary resistivities (although resistivities are in general additive when N-processes dominate, the combination of boundary scattering and dominant N-processes does not follow this rule (see § 7.3)). Anderson and Smith (1973) found striking evidence of this non-additivity by comparing the thermal conductivities of rough and smooth crystals containing defects.

Detailed examination of the Casimir expression shows that the averaging of phonon velocities almost completely masks the anisotropy in

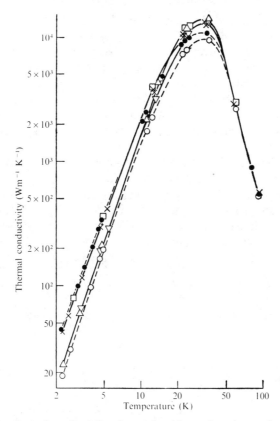

FIG. 7.12. The thermal conductivity of crystals with rough and smooth surfaces. Below ~10 K the six specimens divide into two groups according to the surface finish. Near the maximum they divide into three groups corresponding to the production batch (and so, presumably, to the specimen perfection). (After Berman et al. 1955.)

phonon velocities, so that the boundary scattering-limited 'conductivity' should be very nearly isotropic for most crystals. McCurdy, Maris, and Elbaum (1970) have, however, shown that even in cubic crystals the conductivity limited by boundary scattering can be anisotropic. This anisotropy arises because the group velocity of phonons is not, in general, collinear with the phase velocity, and this results in a bunching of the energy flow in certain directions. In a long crystal of silicon, for example, conductivities in the ratio of 1·8:1 should be observed for suitably different orientations of the rod relative to the crystal axes. The anisotropy and the conductivities themselves are reduced for short crystals, but differences of 1·6:1 were observed, in good agreement with calculation from the known elastic constants. Even in cubic crystals the anisotropy is not necessarily in a particular sense, and in calcium fluoride the directions of high and low conductivity are reversed relative to silicon, again in agreement with the calculations.

Although the discussion which has been given would suggest that boundary scattering is only a low-temperature phenomenon, its effects can, in fact, be observed at temperatures much higher than that of the conductivity maximum. For most scattering processes which take place in the bulk of a crystal, phonons of small wave-number (long wavelength) are much less scattered than those with large wave-numbers. Boundary scattering can then be important in limiting the mean free paths of small-q phonons at temperatures where large-q phonons have their mean free paths determined by the internal scattering. Herring (1954) predicted the size dependence of the conductivity for different bulk scattering laws and Geballe and Hull (1955) found a size dependence in relatively thick germanium specimens (cross-sections 2·1 and 4·1 mm square) at temperatures up to ~10 times the temperature of the maximum.

Savvides and Goldsmid (1972), working with silicon, obtained much

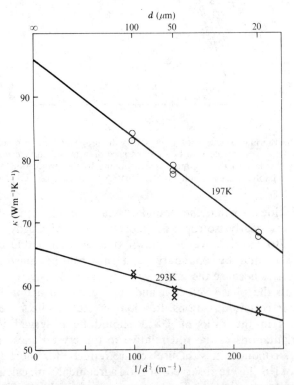

FIG. 7.13. The thermal conductivity of neutron-irradiated silicon as a function of the reciprocal of the square root of the thickness. (From Savvides and Goldsmid 1973.)

more pronounced effects by using thinner specimens and reducing the contribution to the conductivity from phonons of large q even further by irradiating the specimens with neutrons; the relatively small defects produced scatter short-wavelength phonons preferentially. By increasing the accuracy of measurement, they later (Savvides and Goldsmid 1974) detected a small size effect in unirradiated material. Herring had shown that if the dominant bulk scattering law can be represented as $(\tau(q))^{-1} \propto q^2$, then the conductivity at a particular temperature would depend on size according to a relation of the form $\kappa = A - Bd^{-\frac{1}{2}}$. Savvides and Goldsmid (1973) assumed that the bulk scattering of the phonons which are unaffected by the neutron-produced defects was due to N- and U-processes, both scattering so that $(\tau(q))^{-1} \propto q^2$ at fixed temperature. They also found that at a given temperature the conductivity should be linear in the inverse of the square root of the specimen thickness, and their experimental results fitted such a relation up to ~300 K (about 14 times the temperature of the conductivity maximum of pure silicon), as is shown in Fig. 7.13.

7.3. Poiseuille flow of phonons

When discussing boundary scattering, an analogy was drawn with the flow of gas at very low pressure. The flow of heat in a crystal at normal temperature does not have a true analogy in gas flow as there is no equivalent of U-processes in the collisions between molecules. This would require the combined momentum of the interacting molecules to be reversed at a collision. In a real gas at normal pressure the flow is determined by two factors: the mean drift velocity of the molecules in the layer adjacent to the walls and the extent to which this velocity determines the velocity of all other layers of gas. In order to obtain the simple Poiseuille relation (eqn (7.7)) it is assumed that there is no 'slip' and the molecules at the walls have zero drift velocity. The variation of velocity with distance from the walls is determined by the viscosity, which is a measure of the mean free path between the intermolecular collisions. On average, energy is conserved in these collisions and so is momentum, so that the flow is determined by the mean free path for momentum-*conserving* collisions. For heat flow in a crystal to be analogous to this kind of gas flow, conditions must be such that phonons only interact with one another by N-processes and are scattered by the crystal boundaries as well. If we take the essential factors from the Poiseuille expression applied to a gas (eqn (7.7)), we see that the flow is proportional to r^4/η which is proportional to r^4/l, where l is the mean free path for collisions in which momentum as well as energy are conserved. For phonons the corresponding mean free path is that appropriate to N-processes, l_N. If we divide the flow by the cross-sectional area to obtain the 'conductivity', we find that it is proportional to r^2/l_N. We therefore find the perhaps surprising result that whereas the true conductivity at normal temperatures is proportional to the mean free path for processes which do not conserve wave-vector, it is inversely proportional to the mean free path for N-processes, in which the wave-vector is conserved, if these are the only scattering processes which the phonons undergo apart from boundary scattering. If we compare the 'conductivity' under this Poiseuille regime with that when there is pure boundary scattering, as discussed in § 7.2, we see that it is enhanced by $\sim r/l_N$ (exact calculation leads to the factor $\frac{5}{16}r/l_N$ (Sussman and Thellung 1963, Guyer and Krumhansl 1966)).

We can also arrive at the qualitative result by considering the phonons to cross the crystal by a series of random-walk steps between momentum-conserving collisions. If each step is of length l_N, then the number of steps required to travel a distance D is $\sim(D/l_N)^2$. The effective mean free path is thus $\sim(D/l_N)^2 l_N$. For pure boundary scattering we assume an effective mean free path of D, so that the enhancement again comes to $\sim D/L_N$.

Klemens (1958) gave a qualitative prediction of Poiseuille flow with this enhancement of the 'conductivity', but did not deduce the stringent conditions which would need to be fulfilled for its observation. These conditions were given by Gurzhi (1964) and by Guyer and Krumhansl (1966). Scattering by resistive processes must be negligible and there must be many N-processes between interactions of the phonons with the walls. The second condition leads to the requirement $l_N \ll D$. The simple consequence, $l_R \gg l_N$, of the first condition, is not sufficiently stringent and l_R must in fact be so great that $l_R l_N \gg D^2$.

Poiseuille flow of phonons has been most clearly observed in helium crystals, and careful growth at constant pressure produces crystals which are usually perfect enough to show it. Helium is such an anharmonic crystal that N-processes can still persist at low temperatures where the condition for U-processes can no longer be fulfilled. As l_N increases with decreasing temperature, the conductivity, when governed by Poiseuille flow, varies even faster with temperature than the usual T^3 variation of boundary scattering. It cannot be observed over more than a very narrow temperature range because l_N rapidly becomes too large and ordinary boundary scattering takes over.

Mezhov-Deglin (1964) first observed the characteristic rapid drop in conductivity just below the maximum for helium, and the phenomenon has since then been studied in more detail by him (Mezhov-Deglin 1965, 1967) and by Hogan et al. (1969) and Lawson and Fairbank (1973). The enhancement of the heat flow is most readily seen by removing the T^3 variation of the specific heat and plotting the effective mean free path as a function of temperature. At 'high' temperatures the mean free path is

PERFECT NON-METALLIC CRYSTALS

always below D and increases with decreasing temperature. If Poiseuille flow occurs, the effective mean free path actually rises above the value of D but finally decreases again to the value of D, which is characteristic of pure boundary scattering. Fig. 7.14 shows an example of this behaviour of the mean free path in helium crystals. Mezhov-Deglin found that for crystals grown at 85 atm Poiseuille flow could not be observed in specimens with diameter less than 1·3 mm.

Poiseuille flow of phonons was actually first observed by Whitworth (1958) in liquid helium. At sufficiently low temperatures, thermal conduction in the liquid is also due to phonons. From the analogy with gas flow, Whitworth deduced that the enhancement of the heat flow would be $\frac{3}{16}r/l_N$. He was able to deduce a temperature variation for l_N.

The values of l_N and its temperature variation in the solid, deduced from Poiseuille-flow measurements, have been compared with values which can be deduced by analysing other thermal-conductivity measurements (see § 8.3.1(a)) and by analysing experiments on 'second sound'. The agreement is not unreasonable considering that the averages of

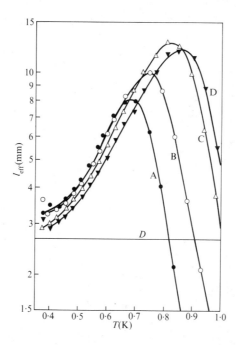

FIG. 7.14. The effective mean free path, $l_{\text{eff}} = 3\kappa/Cv$, in helium crystals grown at 85 atm for four different orientations showing the enhancement above the specimen diameter D characteristic of Poiseuille flow. Angles between specimen axis and crystal c-axis: A, 14°; B, 61°; C, 82°; D, 90°. (After Hogan et al. 1969.)

phonon properties which affect these different phenomena are not the same.

Kopylov and Mezhov-Deglin (1971, 1973) observed less dramatic Poiseuille flow in bismuth single crystals. They ascribed the small enhancement in the conductivity to scattering of phonons by charge carriers within the specimen.

8
THE THERMAL CONDUCTIVITY OF IMPERFECT CRYSTALS

FOR reasonably perfect crystals, U-processes dominate the conductivity at high temperatures and boundary scattering dominates at low temperatures. The conductivity is thus most sensitive to the influence of imperfections in the region of the conductivity maximum, where both these unavoidable components are small. If crystals are not grown with great care from very pure materials, however, the defects can have a large influence on the conductivity over a wide temperature range.

An interpretation of the dependence of conductivity on the presence of defects requires a knowledge of how various types of defects scatter phonons, and how this scattering is to be incorporated with the intrinsic scattering mechanisms into an expression for the resulting conductivity.

Scattering of phonons by lattice defects has been treated by numerous authors (e.g. Klemens 1955, Ziman 1960, Carruthers 1961, to mention only a few important contributions) and the main conclusions will be summarized here.

8.1. Phonon scattering by defects

8.1.1. Point defects

A defect which extends over a volume with linear dimensions much smaller than the phonon wavelength can be considered as a point defect. At a temperature T the dominant phonon wavelengths for thermal conduction are of the order of θ/T interatomic distances (the exact value depends on the scattering law), so that if T is very much less than θ, a defect confined to only a few atomic volumes will be small enough to appear as a point defect to the important phonons. A defect which fits this criterion may be a wrong atom on a lattice site substituting for the correct atom, a vacancy at a lattice site, an interstitial atom, or combinations of these. The scattering is then caused by the difference in mass and the difference in bonding between the atoms. There may also be some distortion of the lattice around the imperfection caused by the different volumes of the correct and incorrect atoms.

Although quantum-mechanical methods applied to a discrete lattice must be used to obtain exact expressions for the rate of scattering of phonons by various defects, there are some cases involving long-wavelength phonons in which a reasonable approximation to the true result may be obtained by applying classical methods for continuous elastic media.

Rayleigh (1896) calculated the scattering of sound waves by an obstruction which has all its dimensions small compared with the wavelength, and has a density and rigidity different from that of the surrounding medium. He obtained results for two different cases: scattering by an arbitrarily shaped region when the differences in density and compressibility are small, and scattering by a sphere for any magnitudes of these differences. The former situation leads to an expression for the scattering cross-section more easily related to that obtained for a discrete lattice by perturbation theory.

If the region has volume V and its compressibility and density differ by $\Delta\chi$ and $\Delta\delta$ from the values χ and δ for the surrounding medium, then the cross-section for scattering of sound waves of wavelength λ is

$$\Gamma(\lambda) = \frac{4\pi^3 V^2}{\lambda^4}\left\{\left(\frac{\Delta\chi}{\chi}\right)^2 + \frac{1}{3}\left(\frac{\Delta\delta}{\delta}\right)^2\right\}.$$

For phonons tranverse as well as longitudinal modes must be considered, and this complicates the contribution to the scattering due to the change in elastic constants. However, if we regard the imperfection as consisting of a medium which is perfectly rigid, the three directions of polarization merely treble the scattering, and its cross-section becomes

$$\Gamma(q) = \frac{V^2 q^4}{4\pi}\left(\frac{\Delta\delta}{\delta}\right)^2.$$

This result should hold for a real lattice if the defect consists of an atom which only differs from the correct atom of mass M by having a mass $M + \Delta M$. Then $\Delta\delta = \Delta M/V$ and $\delta = M/a^3$, where a^3 is the volume occupied by an atom in the lattice (in a simple cubic lattice a would be the nearest-neighbour distance). For wavelengths which are considerably greater than the interatomic distance there is generally little dispersion, so that we may replace the wave-number q by ω/v and derive the relaxation rate for n_p point defects per unit volume:

$$\begin{aligned}\frac{1}{\tau_{\Delta M}(\omega)} &= n_p \Gamma(\omega) v \\ &= \frac{c_p}{a^3}\frac{a^6}{4\pi}\left(\frac{\Delta M}{M}\right)^2 \frac{\omega^4}{v^4} v \\ &= \frac{c_p a^3 \omega^4}{4\pi v^3}\left(\frac{\Delta M}{M}\right)^2. \end{aligned} \qquad (8.1)$$

where c_p is the ratio of the number of point defects to the number of lattice sites per unit volume.

Klemens (1955) obtained the same result for the scattering due to a mass difference using perturbation theory. The perturbation in energy is $\frac{1}{2}\Delta MR^2$, where R is the atomic displacement. The rate of change of occupation number of phonons ω contains the usual resonance factor which ensures that the scattered and incident phonons ω and ω' have the same energy. The matrix element for the transition contains the product $\omega\omega'$, so that the transition probability is proportional to ω^4. The result holds for any ΔM.

A point defect which can approach this ideal of scattering only because of its mass difference corresponds to the insertion of an atom of one isotope into the lattice of another isotope of the element, and we shall see (§ 8.3.1(a)) that experiments on such systems have been useful in investigating the general methods of interpreting thermal-conductivity measurements.

If we do not worry too much about the difference in velocities for waves of various polarizations, the relative change in compressibility can be expressed as twice the relative change in phonon velocity, $2(\Delta v/v)$. The Rayleigh expression for the scattering rate due only to a difference in compressibility between the scattering region and the surrounding medium, again allowing for scattering into three polarization modes, then becomes

$$\frac{1}{\tau_{\Delta v}(\omega)} = \frac{3}{\pi} V^2 q^4 \left(\frac{\Delta v}{v}\right)^2.$$

Again, the expression derived by Klemens is very similar, and if we substitute a^3 for V the only difference is that Klemens's constant is $2/\pi$. In view of the lack of precision in describing a change in binding force between atoms in terms of a change in phonon velocity and the manner in which different polarizations have been lumped together, we should not be surprised that simple modification of Rayleigh's expression does not give exactly the same numerical result as perturbation theory applied to a discrete lattice.

In general, the insertion of a wrong atom into a lattice forces the surrounding atoms into slightly different positions. In a perfectly harmonic lattice the resulting strain does not lead to scattering of phonons, but in a real anharmonic crystal phonon frequencies are changed by changes in interatomic distance and this leads to scattering. The expression for the relaxation rate due to a relative displacement $\Delta R/R$ of the nearest neighbours, given by Klemens, is

$$\frac{1}{\tau_{\Delta R}(\omega)} = \frac{2c_p a^3 \omega^4}{\pi v^3} J^2 \gamma^2 \left(\frac{\Delta R}{R}\right)^2$$

where the constant J depends on how the nearest and further-out linkages combine in the scattering matrix. We may note that the square of the Grüneisen constant again appears where anharmonicity is the cause of scattering.

The sign of the amplitude of the scattered wave depends on the sign of the radial displacement, and, since $1/\tau_{\Delta v}(\omega)$ is also due to changes in atomic linkage, the amplitudes must be added before the scattering intensity is derived. Klemens (1955) wrote the total scattering rate as

$$\frac{1}{\tau(\omega)} = \frac{3c_p a^3}{\pi v^3} S^2 \omega^4 \qquad (8.2)$$

where $S^2 = S_1^2 + (S_2 + S_3)^2$, and S_1, S_2, and S_3 correspond to scattering by differences in mass, velocity (corresponding to compressibility in Rayleigh's formulation), and radial spacing; he tabulated values for the scattering parameters for substitutional atoms and vacancies in alkali halides.

Prompted by some experimental results of Baumann and Pohl (1967), Krumhansl and Matthew (1965) investigated the conditions under which there could be compensating contributions to the scattering from changes in mass and force constant. In order to be able to obtain exact results, they examined the situation in a linear chain in which one particle of mass M is replaced by one of mass $M + \Delta M$. This particle is linked to its two neighbours by harmonic forces with constant $\zeta + \Delta \zeta$. They found that the energy-reflection coefficient in the long-wavelength limit has the Rayleigh frequency dependence (ω^2 for the one-dimensional case) with magnitude proportional to

$$\left\{ \frac{\Delta M}{M} + \frac{2(\Delta \xi / \xi)}{1 + (\Delta \xi / \xi)} \right\}^2.$$

If $\Delta M/M$ and $\Delta \zeta/\zeta$ have opposite signs, partial or even complete cancellation of the scattering amplitude can occur.

Yussouff and Mahanty (1966, 1967) considered the scattering in three-dimensional lattices by mass and force-constant changes. For long-wavelength phonons the total scattering depends on the incident direction, even for cubic lattices, and on polarization. They discussed interference between the two contributions to the scattering, but by numerical evaluation Baumann and Pohl showed that this interference does not, in fact, yield a large enough effect to explain their experimental observations.

8.1.2. Larger defects

If a defect is larger than a few interatomic spacings, the scattering is not proportional to ω^4 for all phonons. Low-frequency phonons are still scattered according to the Rayleigh law, the cross-section being dependent on the volume of the disordered region and not on its shape. At the opposite extreme of wavelengths small compared with all the linear dimensions of the defect, the cross-section depends on the area presented to the incident phonon and should be independent of frequency. Turk and Klemens (1974) found that for a thin circular plate-like defect the scattering cross-section in the short-wavelength limit is proportional to ω^2, and they derived the correction term for intermediate wavelengths, which is dependent on the ratio of wavelength to plate radius, which takes this scattering over to the Rayleigh ω^4 law at long wavelengths.

The cross-section when the phonon wavelengths are comparable with all dimensions of the scatterer cannot be represented by a simple power of the frequency, and, judging by the calculations of Anderson (1950) for sound waves, the variation of scattering with wavelength is very sensitive to the relation between the force-constant and mass changes. For sound waves in a fluid, scattered by a sphere of another fluid, the cross-section can show a smooth transition from Rayleigh to geometrical scattering or pass through a broad maximum or oscillate. However, as thermal conductivity is determined by a wide range of frequencies, we would not expect oscillations in the frequency dependence of the cross-section necessarily to show up as an irregular variation in conductivity. Schwartz and Walker (1967b) have approximated Anderson's results by manageable analytic expressions for a case which gives an oscillating cross-section, and substituted the corresponding relaxation times into the thermal conductivity integral (eqn (4.9b)). The computed conductivity varied with temperature in a manner to be expected if the transition from Rayleigh to geometrical scattering were smooth. The main features of the computed conductivity curves are even reproduced quite well by the more drastic approximation that Rayleigh scattering occurs for wavelengths greater than $2\pi D$, where D is the diameter of the defect, and that for shorter wavelengths the cross-section remains constant at the value appropriate to Rayleigh scattering for a wavelength $2\pi D$.

8.1.3. Dislocations

The scattering of phonons by dislocations has been the subject of considerable theory and experiment. The main features of the scattering by static dislocations can be deduced from classical arguments by considering separately the effects of the core and of the surrounding strain field.

The core of a dislocation consists of a narrow region along its axis, within which there is a drastically altered structure which may be represented by a change in density. The Rayleigh expression for the scattering width for a sound wave due to a rigid and immovable cylinder with radius r, small compared with the wavelength, which has its axis perpendicular to the incident waves can be written

$$\Gamma(\omega) \sim \frac{r^4 \omega^3}{v^3}$$

so that

$$\frac{1}{\tau_{\text{core}}(\omega)} \sim N_D \frac{r^4}{v^2} \omega^3$$

where N_D is the number of dislocation lines per unit area.

The effect of the surrounding strain field can be estimated by making an analogy with geometrical optics. The strain in the crystal at a distance x from the core, where x is not too small, is $\sim B/x$, where B is the Burgers vector of the dislocation (of the order of interatomic distances). Owing to the anharmonicity in real crystals, the strain alters the phonon velocity and this corresponds to a change in refractive index on the optical model, so that the wave is deviated on passing through the strained material. The scattering width is then $\sim \gamma^2 B^2/x_o$, where x_o is the smallest value for which this optical analogy applies, which must be $x_o \sim \lambda$. We thus find

$$\frac{1}{\tau_{\text{str}}(\omega)} \sim N_D \frac{\gamma^2 B^2 \omega}{2\pi}. \tag{8.3}$$

This qualitative result for the scattering by the strain field of a dislocation was derived by Nabarro (1951), and further treatments of the scattering have been given by, among others, Klemens (1955, 1958), Ziman (1960), Carruthers (1961), Bross, Seeger, and Haberkorn (1963), and Ohashi (1968), all of whom obtain similar forms for the scattering. For example, Klemens gives the numerical constant as 6×10^{-2} instead of $1/2\pi (= 1 \cdot 6 \times 10^{-1})$ in the expression for $1/\tau_{\text{str}}$ due to a screw dislocation perpendicular to the temperature gradient, and an additional factor of 0·55 for dislocations arranged at random.

The differences in numerical constants found by different authors are not significant when drawing conclusions about the relative importance of the core and the strain field in scattering phonons. This ratio

$$\frac{1/\tau_{\text{core}}}{1/\tau_{\text{str}}} \sim \frac{r^4 \omega^3/v^2}{\gamma^2 B^2 \omega}.$$

If we assume that the radius of the highly disturbed cylinder is $\sim B$ and that γ is about 1, then the ratio is $\sim B^2/\lambda^2$; B itself is of the order of a and phonons which have wavelengths as small as $\sim a$ have frequencies near the Debye maximum ω_D. Thus, except for the highest frequencies which can propagate, the scattering by the strain field outweighs that due to the core.

A number of experimental studies of the effect of dislocations on the conductivity of non-metallic crystals (see § 8.3.4) suggest that the actual scattering is often 10^2 to 10^3 times larger than that calculated for fixed dislocations. Estimates have been made of the scattering by vibrating dislocations, and experiments by Anderson and Malinowski (1972) and Suzuki and Suzuki (1972) provide strong evidence that these can account for observed reductions in conductivity. Similar evidence is obtained from measurements on superconductors (see § 12.2.4).

Nabarro (1951) calculated the scattering of sound waves (long shear waves) by a mobile screw dislocation and showed that the scattering width is $\sim \lambda$. Ziman (1960) treated a dislocation as a rigid cylinder which was free to move ('flutter') under the action of the stress field of the incident phonon, and also obtained a scattering width $\sim \lambda$, so that $1/\tau_{\text{flut}}(\omega) \sim 2\pi N_D v^2/\omega$. The ratio

$$\frac{1/\tau_{\text{flut}}}{1/\tau_{\text{str}}} \sim \frac{\lambda^2}{\gamma^2 B^2}$$

which apart from $1/\gamma^4$ is the inverse of $(1/\tau_{\text{core}})/(1/\tau_{\text{str}})$ and is large except for the shortest wavelengths.

Garber and Granato (1970) calculated the scattering on an elastic string model for a dislocation, with the string fixed at pinning points with a distribution of loop lengths. On this model too, with reasonable average lengths, the scattering is greater than for static dislocations, and $1/\tau$ increases with decreasing frequency, roughly as $1/\omega$, reaching a maximum when the wavelength is about 10 times the mean loop length. The scattering then falls off rapidly. Ninomiya (1968) has calculated the scattering by interaction with travelling waves on an infinitely long elastic string. The frequency dependence of this scattering is small, but the magnitude is again larger than that calculated for static dislocations.

8.2. Combining intrinsic and defect scattering rates

If we wish to calculate the conductivity and its temperature variation when known defects are present, or deduce from the measured conductivity how the defects present scatter phonons, a lengthy computation is usually required. Several scattering mechanisms apart from the defects being studied have to be taken into account and the effect of any one mechanism is dependent on the others, so that it is necessary to 'guess' relaxation times which are to be fed into the kinds of expression discussed in Chapter 6. The influence of N-processes makes the task more difficult.

It is, however, sometimes possible to derive some information from much simpler approaches, which will first be mentioned briefly.

8.2.1. Dominant phonon method

In this method it is assumed that the relaxation time to be used in the simple kinetic expression for heat conductivity, $\kappa = \frac{1}{3}Cv^2\tau$, is just the relaxation time for the phonons which are dominant in carrying the heat. If one type of defect is mainly responsible for the scattering, which can be described by a relaxation rate $1/\tau \propto \omega^z$, then from the definition of x ($= \hbar\omega/k_B T$) $1/\tau \propto x^z T^z$. If the temperature is sufficiently low, the phonons dominant in heat conduction correspond to x_{dom} being constant, so that $\omega_{\text{dom}} \propto T$ (the constant depends on the scattering law). Taking $C \propto T^3$, we then have $\kappa \propto T^3 T^{-z} \propto T^{3-z}$. This implies, therefore, that if point defects ($1/\tau \propto \omega^4$) are the dominant scatterers at low temperatures then $\kappa \propto T^{-1}$, or that if sessile dislocations ($1/\tau \propto \omega$) are dominant then $\kappa \propto T^2$, and these are fair representations of experimental observations.

At high temperatures the dominant phonons cannot have ever-increasing frequency, since there is a limiting frequency $\omega_D = (k_B \theta)/\hbar$. On our simple model ω_{dom} is constant at high temperatures at this limiting value. Since the specific heat is also effectively constant at these temperatures, the high-temperature conductivity, when dominated by defect scattering, should be independent of temperature.

We can dress these results in a little more respectability by writing eqn (4.9b), with the constant factor represented by I, as

$$\kappa = IT^3 \int_0^{\theta/T} \tau(x) \frac{x^4 e^x}{(e^x - 1)^2} dx.$$

For $1/\tau \propto \omega^z$ we obtain

$$\kappa = I'T^{3-z} \int_0^{\theta/T} \frac{x^{4-z} e^x}{(e^x - 1)^2} dx.$$

If θ/T is large enough, because the temperature is low, the integral is a number independent of the upper limit of integration, so that the conductivity is proportional to T^{3-z}. At the opposite extreme of high temperature, x is always small. For small x, $(e^x - 1)^2 \sim x^2$ and $e^x \sim 1$, so that

$$K = I''T^{3-z} \int_0^{\theta/T} x^{2-z} dx$$

which is independent of T (unless $z = 3$).

In these derivations the divergence which occurs for $z > 2$ has been ignored. In actual crystals other scattering processes, such as N- and U-processes and boundary scattering, prevent the phonons with small values of x from running away with the heat flow. As a result the effective lower limit for the integration is not zero, and the slope of a conductivity curve at low temperatures often provides a reasonably reliable guide to the frequency dependence of defect scattering and thus to the nature of the defects. However, we would not expect such deductions to be possible from measurements at high temperatures.

As an example of this method we may consider the effect of F-centres on the conductivity of potassium chloride. Walker (1963) found that, for a particular specimen of KCl containing F-centres produced by additive colouration, the conductivity in the region of 1 K was proportional to T^n, where n was a little more than 2·6. This is slightly steeper than the curve for the untreated crystal and suggests that, in the absence of other scattering mechanisms, the conductivity determined by F-centres alone would be roughly proportional to T^3, corresponding to a relaxation rate independent of frequency. Detailed analysis of the results by the methods to be described did, in fact, lead to the result that $1/\tau$ for these F-centres was independent of ω. Devyatkova and Stil'bans (1952) found that the additional heat resistance introduced in KCl by F-centres produced by additive colouration was nearly the same at 98, 200, and 298 K. This is just what would be expected for high temperatures, regardless of the scattering law for the defects.

8.2.2. The 'Debye approximation'

A much more productive method for analysing experiments made at low temperatures, which was used to great effect by Pohl and his co-workers (see, for example, Walker and Pohl 1963), consists in the direct application of eqn (4.9b). The measured thermal conductivity and its temperature variation are fitted by trial and error for one crystal, representing each scattering mechanism by a suitable relaxation rate, the relaxation rate for each frequency being the sum of relaxation rates appropriate to each mechanism supposed to be acting. The conductivity of a crystal containing extra defects is then fitted by choosing a suitable relaxation rate to represent the scattering by the additional defects. Many interesting facts about defects have been unravelled by finding the appropriate scattering powers. This method has been called the Debye approximation, because in essentials one is just applying the Debye expression for specific heats to provide a more refined form (eqn (4.11)) for the conductivity than the simple kinetic expression (eqn (3.5)).

The Debye approximation only works well because even pure crystals are usually just not perfect enough for N-processes to be very important. In a highly perfect crystal, when N-processes may be dominant, this simple relaxation-time method must be modified. In Callaway's method, for example, the conductivity then consists of two terms (eqn (6.1)) and simple addition of relaxation rates, even within an integral, does not give a proper representation of the conductivity. However, if even in the 'pure' crystal resistive scattering is great enough compared with scattering by N-processes, we may ignore the relaxation rate due to N-processes and the second Callaway term becomes negligible. At the same time the first term is just the same as eqn (4.9b), because N-processes have been neglected.

Much of the work carried out by Pohl was on alkali halides. Many possible constituent elements, such as lithium, potassium, chlorine, and bromine, consist of mixtures of isotopes in appreciable proportions, so that even a chemically pure crystal contains the source of much resistive scattering. The conductivity of the 'pure' crystal can thus be represented quite well by the simple Debye expression, and the conductivity of specimens containing intentional additional defects can be represented in the same way merely by adding to the relaxation rate in the appropriate way.

It has been imperative to use the full Callaway expression to interpret experiments on the effect of isotopes when the initial pure crystal was almost isotopically pure. The addition of the first small concentration of isotopes reduces the conductivity in a much more dramatic way than would be expected from the behaviour of the single-term expression. This occurs because the second Callaway term, κ_2, is large for a pure crystal but decreases very rapidly with addition of defects, so that it is dominant over a wide temperature range for the pure crystal and becomes almost negligible for an impure crystal. The first term, κ_1, varies much more slowly with defect concentration.

A full Callaway analysis gives an indication of the relaxation rate for N-processes too, again because the rapid decrease in κ_2 is very sensitive to their relative importance.

8.2.3. *The variational method*

There have also been analyses of conductivity by the variational method. In general these analyses are very laborious and cannot be included under the heading of simple methods. In one series of experiments on helium, however, it was possible to assume that N-processes dominated the phonon distribution even when defects had been introduced into the crystals (see § 8.3.1(a)). As has been shown in §§ 6.1.2 and 6.2.2, both the variational method and the Callaway method give the same result that resistances are strictly additive in this situation. This particular analysis can thus be described as a simple application of either the Callaway or the variational method.

8.3. Thermal conductivity of crystals containing defects

No attempt will be made to describe more than a very small number of the measurements and analyses which have been made. Much of the work has been carried out at moderately low temperatures, since the temperature dependence of the conductivity is then an important diagnostic element. At too high a temperature, defects add a resistivity which is little dependent on temperature, while at too low a temperature the important phonon wavelengths are so long that scattering by most defects becomes negligible. Some defects scatter phonons resonantly, so that their influence does not change monotonically with temperature.

8.3.1. Non-resonant point defects

8.3.1(a). Isotopes. An isotope of an element at a site in the lattice composed of another isotope of the same element constitutes the simplest type of point defect. The calculation of the scattering if the crystal consists of more than one element, one or more of which have isotopes, is hardly more difficult. At the temperature at which the crystal is grown the distribution of isotopes is random, and, except for helium, ordering does not occur during the duration of an experiment. For a 'classical' crystal we expect no change in the crystal other than a difference of mass of the atom at the site where the isotope occurs. However, in crystals whose properties are governed by quantum effects, of which helium is the most notable example, we can expect lattice distortions and changes in effective force constants in the neighbourhood of the isotopic defect.

Detailed studies have been made of the effect of isotopic constitution on the conductivity at fairly low temperatures for lithium fluoride (^{6}LiF–^{7}LiF), helium (^{3}He–^{4}He), and neon (^{20}Ne–^{22}Ne). In addition, the conductivity of a germanium crystal made from material in which ^{74}Ge had been enriched to 95 per cent concentration was compared with that of crystals made from natural germanium.

All these measurements have been analysed by the Callaway method (eqn (6.1)), and in fact the results of Geballe and Hull (1958) on germanium were analysed in the paper which originally expounded the method (Callaway 1959). Although the fit to the germanium data was satisfactory, it is more convincing if there is a whole set of curves for different isotopic concentrations which can be fitted by varying only that relaxation rate which is ascribed in the analysis to the isotopes. Such an analysis was carried out for lithium fluoride by Berman and Brock (1965) who measured crystals which ranged from nearly pure ^{6}LiF to nearly pure ^{7}LiF.

Fig. 8.1 shows the experimental results for LiF and the best fit which could be obtained by the Callaway method. The theoretical relaxation rate for boundary scattering, assuming perfectly rough crystal surfaces, can be calculated from the cross-sectional dimensions and the mean phonon velocity, and the experimental value can be deduced from the conductivity at the lowest temperatures. The difference between the two was small. Isotopes and N-processes become important at higher temperatures and the best relaxation times for these have to be judged from the

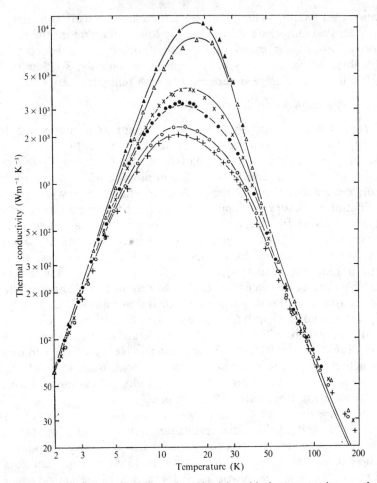

FIG. 8.1. Comparison of the experimental conductivity with the computations on the full Callaway model for LiF. The ^{6}Li concentrations are as follows (per cent): ▲ 0·02; △ 0·01; × 4·6; ● 90·4; ○ 25·0; + 50·1. It was necessary to assume the presence of other impurities in the first two specimens. Two other specimens are omitted because their conductivities agree closely with two of those shown: 7·4 with 90·4 and 75·7 with 25·0 per cent. (After Berman and Brock 1965.)

combination which reproduces best the shapes of the experimental curves and the spacing between them. This is rather subjective, but for LiF it was possible to show that, if isotope scattering were represented by the classical Rayleigh expression (eqn (8.1)), then the relaxation time for N-processes must obey a law $\tau_N^{-1} \propto xT^\Omega$, where Ω is between 4 and 5 (equivalent to $\omega T^{\Omega-1}$). It was not possible to vary the isotope scattering rate to be very different from the Rayleigh value and still fit the experimental results. At still higher temperatures U-processes become important, but as discussed in § 7.1.2(a) it was not possible to decide a unique form for the U-process relaxation rate, although its magnitude could be fixed fairly well. The combined relaxation rate used in the Callaway expression which gave the fit shown in Fig. 8.1 was

$$\tau_C^{-1} = \tau_N^{-1} + \tau_R^{-1}$$
$$= 35xT^4 + 1\cdot 04 \times 10^6 + d'xT + \alpha x^4 T^4 + 600 x^2 T^4 \exp(-170/T) \; s^{-1}.$$

The individual rates represent scattering by N-processes, boundaries, dislocations (a small term which produced some improvement in the fit and could be related to the etch-pit count), isotopes (the Rayleigh value), and U-processes. In terms of frequency, the rate for N-processes is $2\cdot 7 \times 10^{-10} \omega T^3 \, s^{-1}$.

The contributions of the two Callaway terms κ_1 and κ_2 (eqn (6.1)) are shown in Fig. 8.2 for the relaxation rates used in the analysis. It can be seen that κ_2 decreases much more rapidly than κ_1 with increasing isotope concentration. An isotopically pure crystal would correspond to $\alpha = 0$, while 50 per cent of one isotope corresponds to $\alpha = 1\cdot 2$.

There is indirect support for the correctness of the isotope scattering rate, apart from the fit obtained in these experiments. Using similar methods of analysis (see below) it was not possible to use the appropriate Rayleigh values for helium and neon, which are 'quantum crystals' in which an isotope is expected to have a more drastic effect on the lattice than in LiF. This suggests that when the scattering is different from the Rayleigh prediction, this is found out by the analysis.

There is some confirmation of the form of the N-process relaxation rate used, derived from measurements of ultrasonic attenuation in lithium fluoride crystals by de Klerk and Klemens (1966). The conductivity curves are most sensitive to the form of the N-process relaxation rate in the range from 10 to 40 K, where the frequencies of the dominant phonons are $\sim 10^{12}$ Hz. The ultrasonic measurements were made over a similar temperature range but at frequencies of 4×10^8 and 10^9 Hz. For the slow transverse mode the ultrasonic attenuation was found to be proportional to $\omega T^{3\cdot 5}$ and at 15 K the rate was $2 \times 10^{-6} \omega \, s^{-1}$, while from the thermal-conductivity analysis the temperature variation was similar and the value at 15 K was just under $1 \times 10^{-6} \omega \, s^{-1}$. This agreement does have some significance, since on a simple picture the slowest modes should provide the greatest contribution to the conductivity because the relevant velocity occurs in the denominator of an expression such as eqn (6.1). Moreover, at 15 K in a substance with a Debye temperature of over 700 K there would be little dispersion at the frequencies important in thermal conduction, and the continuity in the properties of the phonons

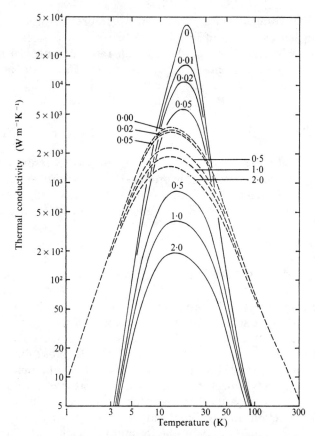

FIG. 8.2. Computed curves for κ_1 and κ_2 on the Callaway model as a function of isotope scattering for LiF. Full curves are κ_2 and broken curves κ_1. The number beside each curve is the corresponding value of α (in $s^{-1} K^{-4}$), the coefficient of the term in $x^4 T^4$ in the phonon-scattering rate. (After Berman and Brock 1965.)

of a particular branch would be expected to make extrapolation to lower frequencies meaningful.

A similar set of measurements accompanied by a Callaway-type analysis has been carried out on neon by Kimber and Rogers (1973). Again, the fit of computed curves to the experimental results for the whole range of isotopic concentrations is very sensitive to the relaxation rate ascribed to N-processes, but for this case the best fit was not for an isotope scattering rate exactly as given by eqn (8.1). The enhancement of the scattering, given by the ratio $(1/\tau_{\text{exp}})/(1/\tau_{\Delta M})$, where the value of the numerator is determined by the best fit to the experiments and the denominator is the mass-only value given by eqn (8.1), lay between 1·25 and 1·65. Outside these limits the fit to the experimental results was definitely impaired, however much the form of $1/\tau_N$ was altered. It was concluded that an isotopic impurity in neon definitely does more to the lattice than just introduce a change in mass at the lattice site it occupies. Jones (1970) calculated the extra scattering above the purely mass effect to be expected for neon isotopes due to the long-range strain field around an isotopic impurity. Depending on the assumptions made, the calculated enhancement of the scattering rate was between 1·2 and 2·2, with an intermediate value of 1·4.

Helium is the extreme quantum solid in which the properties of the crystal are governed by quantum effects. The zero-point energy is so high and the van der Waals forces so small that helium only solidifies when under pressure. When the atoms are forced closer together the van der Waals forces increase more rapidly than the zero-point energy and eventually hold the atoms together as a crystal. The zero-point energy is inversely proportional to the atomic mass, so that one isotope in a crystal composed predominantly of the other isotope can be expected to affect the lattice rather profoundly.

It is possible to form helium crystals of very different densities, and at high densities the crystals become more 'classical' in their properties. The enhancement of the isotope scattering would then be expected to decrease with increasing crystal density.

There have been a number of experiments to determine the effect of isotopic constitution on the conductivity of helium, and again the relaxation rates assumed for isotope scattering and for normal processes are interconnected in the analysis. The measurements of Berman, Bounds, and Rogers (1965) and of Berman, Bounds, Day, and Sample (1968) on ^{3}He in ^{4}He cover a density range of 2 and a range of values for Debye θ of 4. From a Callaway analysis they deduced that the enhancement of the relaxation rate decreased steadily from about ~3 for the lowest-density solid to ~1·0 at the highest density. The essential correctness of the analysis was confirmed by Lawson and Fairbank (1973) who introduced extremely small amounts of ^{3}He into the ^{4}He used to form the crystals (10 and 15 ppm). With this very small concentration of isotopes normal processes are so dominant in determining the phonon distribution that it can be assumed that thermal resistances are additive according to eqn (6.4a). It is, therefore, possible to subtract the resistance of the pure crystal from that of the crystals containing the isotopic impurity in order to derive the isotope scattering immediately. Because of the anisotropy in conductivity (§ 7.1.2(b)) the crystals had to be grown in the same orientation, and Fig. 8.3 shows that even for the extremely small ^{3}He concentrations used, accurate resistivity differences could then be derived. The value of $\tau_{\Delta M}(\omega)$ from eqn (8.1) substituted into eqn (6.4a) leads, for the crystal density used, to a mass-difference scattering resistance of

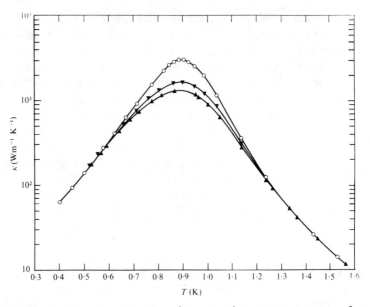

FIG. 8.3. The thermal conductivity of pure ^{4}He and of ^{4}He with 10 and 15 ppm ^{3}He. The crystals were grown at 85 atm pressure with their length perpendicular to the c-axis. (After Lawson and Fairbank 1973.)

11·65 $c_3 T$ m K W^{-1}, where c_3 is the ^{3}He concentration. The experimentally determined extra resistance between ~0·7 and 1·05 K was found to be 31 $c_3 T$ m K W^{-1}, representing an enhancement of 2·7. At the same density the enhancement found by Berman et al. (1965) was ~1·9. In view of the complexity and somewhat subjective nature of a Callaway analysis, this may be considered to be very good agreement between the two very different ways of finding the enhancement.

Since N-processes were dominant under the conditions of the experiments of Lawson and Fairbank, the relaxation rate for N-processes cannot be derived from their analysis. There are, however, other methods of estimating τ_N with which the analyses of Berman et al. (1965, 1968) can be compared. The strength of the N-process scattering can be inferred directly from both the enhancement of the boundary scattering under conditions of Poiseuille flow (see § 7.3) and from the propagation properties of second sound (a wave motion in which the oscillating property is the phonon density). From such experiments and from the analysis of thermal conductivity, τ_N can be expressed as functions of θ_0/T, where θ_0 is the value of the Debye temperature appropriate to the specific heat near absolute zero. The second-sound and Poiseuille-flow experiments suggest that τ_N is of the order of $10^{-12}(\theta_0/T)^3$ s for the dominant phonons, while from heat conductivity it appears that the power of θ_0/T is between 4 and 5, with a correspondingly smaller constant.

If we take a temperature of $\theta_0/30$, which is on the low side of the temperature range for which the conductivity is very sensitive to τ_N, and on the high side of the range in which second sound and Poiseuille flow can be analysed, then the absolute magnitudes of τ_N are similar. The Poiseuille-flow results (Lawson and Fairbank 1973, Hogan *et al.* 1969, Goulder 1973) are consistently one to five times greater than the values derived from second sound (Ackerman and Guyer 1968, Mueller and Fairbank 1971, Fox 1971), which tend to be slightly higher than the thermal-conductivity values. In view of the complexity of the various analyses and the different averages of τ_N probed by different physical processes, this agreement to an order of magnitude should be considered a satisfactory indication that the theory of the role of N-processes in thermal conductivity as represented in the Callaway method is a good approximation to the true situation.

The variational method was applied by Sheard and Ziman (Berman *et al.* 1959) to estimate the thermal conductivity under conditions when only N-processes and point defects are important, and an outline of their method was given in § 6.2.1. The assumption of these conditions restricts the application of the calculations to a very narrow temperature region for any given material. The results of the calculation were expressed in terms of the ratio of the computed heat conductivity to the value it would have for the same defect concentration if N-processes dominated the phonon distribution and point defects were the only source of resistive scattering. This ratio is 1 if defect scattering is very weak, and increases indefinitely as the defects come to dominate the conductivity. Of course, the conductivity itself does not increase as the number of defects increases, but it decreases less rapidly than it would if N-processes remained dominant for all defect concentrations. Comparison was made with experimentally determined conductivities for quite a wide variety of crystals containing point defects, both foreign atoms and, in the case of lithium fluoride, isotopes. The analysis was only expected to be valid at temperatures of about $\theta/20$, and at such temperatures there was quite good agreement.

A variational calculation which includes U-processes and boundary scattering as well as N-processes and mass-difference scattering has been given for germanium by Hamilton and Parrott (1969). In order to render the calculation amenable to numerical computation, they had to make some simplifying assumptions; they used essentially a continuum model with no dispersion and assumed that the crystal was elastically isotropic. U-processes had to be grafted onto the model because they would not occur in a true continuum. However, the intrinsic scattering transition probabilities were derived from measured elastic constants, and the results obtained without adjustable constants were in good agreement with the experiments on natural and isotopically enriched germanium of Geballe and Hull (1958) and on natural germanium of Slack and Glassbrenner (1960) between 2 and 300 K. According to the analysis of Hamilton and Parrott, between 80 and 90 per cent of the heat transport can be ascribed to transverse phonons over all this temperature range.

8.3.1(b). Other point defects. The first low-temperature experiments on the effect of adding known amounts of impurity, with a deduction of the scattering strength from an analysis of the conductivity, were probably those of Slack (1957). He measured KCl crystals containing various concentrations of calcium impurity. The defects consisted of Ca^{2+} ions replacing K^+ ions, with an equal number of positive-ion vacancies to maintain electrical neutrality. Slack's analysis was made according to Klemens's (1951) theory mentioned in § 6.2.1. The change in conductivity on introducing the impurities could not be explained completely by Rayleigh scattering, and Slack concluded that some of the calcium had precipitated to form clusters. He could, however, deduce the coefficient of that part of the scattering rate which had an ω^4 dependence. In terms of Klemens's factor S^2 (eqn (8.2)), the combined S^2 to fit the experiments was 1.4. Klemens computed S^2 for a Ca^{2+} impurity in KCl to be above 0·4 (the uncertainty in the value derives from uncertainty in the effective velocity change) and for a K^+ vacancy to be 0·8. If the impurity and the vacancy are quite separate the two contributions to S^2 should be added. However, if the two are associated, S^2 becomes very small because the velocity and radial contributions from the vacancy are of opposite sign to the large radial distortion due to the Ca^{2+} impurity, and these have to be added as $S_1 + S_2$ before being squared and added to the mass-difference contribution. The resulting scattering was expected to be rather small and to be represented by a value of S^2 less than 0·2. More recent experiments on the same system by Walker and Pohl (1963), which were analysed with the Debye-type integral and included a resonance-scattering term, gave only a slightly smaller value of S^2 (0·9).

Analysis of similar measurements on KCl crystals containing Na^+ and I^- impurities by Walker and Pohl also show only moderate agreement between the derived point-defect scattering and Klemens's calculated values. One should not expect exact agreement since it is difficult to compute the changes of bonding and the lattice distortion introduced by an impurity. The main conclusion is that the scattering can be very different from that due to mass change alone; for a much heavier impurity, such as iodine substituting for chlorine the scattering is only 1·4 times the mass effect, but for Ca^{2+} substituting for K^+ with almost the same mass the scattering is about 80 times the mass effect.

It was shown in § 8.2.1 that on a dominant-phonon model we would expect the resistance introduced by defects at high temperatures to be independent of temperature, whatever the scattering law. This result came from a gross oversimplification of the theory, and experiments by Abeles, Beers, Cody, and Dismukes (1962) on silicon–germanium alloys show that the extra thermal resistance produced by adding germanium to silicon increases slowly with temperature between 300 and 900 K. Abeles (1963) analysed these experiments in terms of an expression he derived for the ratio of the resistance of an alloy to the resistance of pure silicon at the same temperature. He used Callaway's expression for the conductivity in the presence of N-processes and assumed that, at the temperatures concerned, both N- and U-processes had relaxation rates proportional to ω^2 with a fixed ratio between their magnitudes. He also assumed that the point-defect scattering was given in a simple way by the mass and size differences between the two atoms. The adjustable constants in making the fit to the experimental results are then the ratio of the N- and U-process relaxation rates and the effective value of the Grüneisen constant γ. The agreement between the experimental and computed values, for a particular choice of these parameters ($\tau_N^{-1}/\tau_U^{-1} = 2\cdot 5$, $\gamma = 1\cdot 77$) is shown in Fig. 8.4.

8.3.2. Resonant scattering

8.3.2(a). Non-magnetic defects. The first report of the necessity of including a relaxation rate with a resonance character in the analysis of

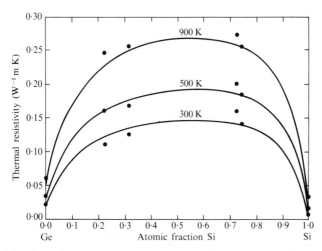

FIG. 8.4. The thermal resistivity of Ge–Si alloys as a function of composition for three different temperatures. The solid curves were computed taking into account both N- and U-processes. (After Abeles 1963.)

the thermal conductivity of crystals containing defects followed measurements on KCl crystals containing small amounts of KNO_2 (Pohl 1962). The conductivity curves for the crystals with defects showed dips between 5 and 10 K, as shown in Fig. 8.5 taken from some later work. An additional relaxation rate proportional to the defect concentration and of the form $\omega^2/(\omega_0^2 - \omega^2)^2$, where ω_0 is some resonance frequency, was found to account for the dips quite well. A damping term in the denominator was not necessary, because the computed conductivity curves are not sensitive to the difference between the large but always finite scattering rate which this would ensure, and the infinite rate at resonance which occurs without damping. This original work was followed by combined measurements of infrared absorption and heat conductivity on several systems by Seward and Narayanamurti (1966) and by Narayanamurti,

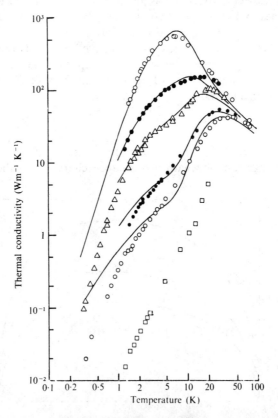

FIG. 8.5. The thermal conductivity of pure KCl and of KCl containing nitrite ions. The solid curves were computed by using a resonant form of the scattering rate. The highest curve is for undoped KCl. The NO_2^- concentrations for the other curves are 9×10^{22}, 4×10^{23}, $1 \cdot 6 \times 10^{24}$, 4×10^{24}, and 4×10^{25} m^{-3}. (After Narayanamurti et al. 1966.)

Seward, and Pohl (1966). As an example of the correlation between these two types of measurements we may take the case of NO_2^- in KCl. In order to fit the conductivity curves the resonant frequency required was $\omega_0 = 3\cdot 8 \times 10^{12}\,\mathrm{s}^{-1}$ $(2000\,\mathrm{m}^{-1})$. From the infrared studies it was deduced that free rotation of the NO_2 ion started at 1200, 1100, and 4000 m^{-1} for three different axes. Phonon spectroscopy by thermal-conductivity measurements is not a very sharp instrument for resolving three such resonances, and it was suggested that the value of 2000 m^{-1} for ω_0 required to fit the conductivity is, in fact, a mean value which takes into account the three resonances found in the infrared work. In other hosts the resonances are associated with tunnelling states, but phonons seem to be weakly coupled to librational states. Similar experiments were also made with CN^- ion impurities.

It may be remarked that the dominant-phonon method allows an order of magnitude to be deduced for resonance frequencies, especially when the associated dips occur below the temperature of the conductivity maximum. If scattering is mainly by the boundaries, the dominant phonons are those which are dominant in the specific heat and correspond to $\hbar\omega/k_B T = x = 3\cdot 8$. The dips for NO_2^- in KCl were centred between 5 and 10 K, and the dominant phonons at 7 K have

$$\omega = 3\cdot 8 \times 7 \frac{k_B}{\hbar} = 3\cdot 5 \times 10^{12}\,\mathrm{s}^{-1},$$

which is very close to the resonant frequency which was required to fit the conductivity from a detailed analysis. This method cannot reveal any of the details of the scattering; one is merely assuming that the maximum of this type of scattering occurs when the dominant phonons are in resonance with the scattering mechanism.

Between the first observations of resonance scattering by molecular impurities and the later work quoted, it had been found that some atomic impurities also changed the conductivity in a way which could only be interpreted in terms of resonance scattering. Walker and Pohl (1963) measured KCl crystals containing Na^+, I^-, and Ca^{2+} impurities. Apart from the Rayleigh type of scattering already discussed, they found that a resonant relaxation rate was also required, but the simple form used by Pohl (1962) was not adequate. They used an expression for the scattering by the impurity mode, given by Wagner (1963), which has its maximum at roughly half the frequency of the impurity mode.

Schwartz and Walker (1966, 1967a) added the divalent ions Ca^{2+}, Sr^{2+}, Eu^{2+}, and Ba^{2+}, which have very different masses and radii, to KCl and observed resonant dips in the conductivity in the region of 30 K for all the impurities. They suggested that the dominant phonon scatterers are the positive-ion vacancies associated with the impurity, and these would be common to all impurities. They were able to fit their results with a relaxation rate due to Krumhansl (1964) describing one-phonon scattering due to the changes in force constant; this has a resonance form with the maximum relaxation rate occurring for a phonon frequency about one-half the Debye maximum frequency. If the dominant phonons are those for which $x = 3.8$, then phonons with $\omega = \frac{1}{2}\omega_D$ would be dominant at $T = \frac{1}{2}\theta/3\cdot8 = 30$ K for KCl.

8.3.2(b). Paramagnetic crystals. Magnetic ions can also act as resonant scatterers of phonons, and their influence on lattice conduction has been investigated intensively since 1960. In ordered magnetic crystals the spin system can conduct heat itself in addition to acting as a source of phonon scattering.

Interactions between spins and phonons give rise to various spin-lattice relaxation processes, and the study of these, from the spin point of view, by essentially magnetic methods is a familiar part of the science of paramagnetism. A spin which is in a higher energy level may relax to a lower level by emitting a phonon of energy $\hbar\omega$ equal to the difference in energy, ΔE, between the two spin states (direct process), or by absorbing one phonon of energy $\hbar\omega_1$ and emitting another of energy $\hbar\omega_2 = \hbar\omega_1 + \Delta E$ (either a Raman or an Orbach process).

These processes can also be looked at from the phonon point of view. A phonon is absorbed on exciting a spin to a higher level by a direct process, and another phonon of the same energy is subsequently emitted in an unrelated direction as the spin returns to its initial state. In indirect processes the phonons which are absorbed and emitted have different energies. All these scattering processes provide a source of thermal resistance.

Slack and Newman (1958) observed a sharp change in slope of the conductivity of MnO at the temperature of the transition from paramagnetic to antiferromagnetic behaviour (the Néel point), which they attributed to interaction between phonons and the magnetic moments of the manganese atoms. However, the forerunner of the extensive studies of such interactions by thermal-conductivity measurements was the work of Rosenberg and Sujak (1960) on zinc sulphate containing Fe^{2+} ions. They found that the conductivity of $ZnSO_4$ was reduced to a far greater extent by the presence of Fe^{2+} impurity than by the same amount of non-magnetic magnesium, although the mass difference between Zn and Mg is four times greater than the difference between Zn and Fe. The influence of a magnetic field on thermal conductivity of crystals containing paramagnetic ions was first examined in experiments by Morton and Rosenberg (1962) on holmium ethyl sulphate and by Dreyfus, Lacaze, and Zadworny (1962) on aluminium oxide containing vanadium and chromium.

Although magnetic energy level systems are rarely so simple in reality, the principles underlying the variation of conductivity with magnetic field can be illustrated by the simple model proposed by Berman, Brock, and Huntley (1963). For Fig. 8.6a it is assumed that the only magnetic levels involved derive from a ground-state doublet which is split by a magnetic field B. The energy difference between the levels is $\Delta E = g_L \beta_B B$, where g_L is the spectroscopic (Landé) splitting factor and β_B is the Bohr magneton. For Fig. 8.6b it is assumed that there is a splitting between the levels even in the absence of a field. Fig. 8.6c represents the case of two inequivalent sets of ions, X and Y, for both of which the splitting is proportional to B, but g_L^X and g_L^Y differ. In Fig. 8.6d there is a singlet level above the doublet, but it is close enough to be excited at the temperatures concerned.

If interactions among phonons are neglected, we can consider that phonons within an energy range $\hbar\,\delta\omega$ make their distinctive contribution

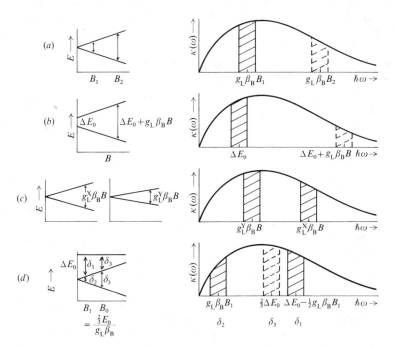

FIG. 8.6. Schematic representation of energy bands removed from the conductivity spectrum by spin interactions for four different idealized level systems. All the bands have been drawn with the same width. (a) Doublet with energy splitting proportional to magnetic field B. (b) Doublet split in zero magnetic field. (c) Two doublets for which the energy splittings are different in general. For the actual case illustrated in Fig. 8.7 (c) these splittings vary with field orientation differently and are equal for certain angles. (d) A singlet lying above a doublet. There are two different splittings to be considered in general (the third and largest is ignored) but these are equal at the particular field B_0.

$\kappa(\omega)\,\delta\omega$ to the thermal conductivity. If their relaxation time in the absence of spin scattering is $\tau(\omega, T)$, then on the Debye model

$$\kappa(\omega)\,\delta\omega \propto \frac{\omega^4 \exp(\hbar\omega/k_B T)\tau(\omega, T)}{T^2\{\exp(\hbar\omega/k_B T) - 1\}^2}\,\delta\omega.$$

The curves on the right-hand side of Fig. 8.6 represent the variation of $\kappa(\omega)$ with $\hbar\omega$.

For the field B_1 the spins in Fig. 8.6a are 'on speaking terms' via a direct process with phonons of energy $\hbar\omega_1 = \Delta E_1 = g_L\beta_B B_1$. The contribution of these phonons in the absence of this interaction is indicated by the ordinate at this energy. In real crystals the energy levels are not sharp, as is suggested in the left-hand diagrams, and a band of phonons can interact with the spin system; such a band is represented by the width of the shading in the right-hand diagrams. For our simple model we assume that the scattering is so violent that there is effectively no contribution to the conductivity from the phonons which interact with the spins, and the reduction in conductivity is then represented by the shaded area (the fractional reduction is the ratio of the shaded area to the total). At a fixed temperature the position of the band varies with B, and the reduction in conductivity is clearly greater at B_1 than at B_2. The maximum reduction occurs for a field such that the band is centred under the maximum.

In Fig. 8.6b a band of phonons is scattered even in zero field so that the conductivity is lower than it would be in the absence of spin scattering. For large enough fields, however, the band which is scattered was originally making such a small contribution that its loss is unimportant and the conductivity becomes larger than it was in the absence of a field.

In Fig. 8.6c two bands of phonons are scattered in general, but for the practical case to be described g_L^X and g_L^Y vary with the angle between the field and the crystal axis, and for certain angles they are equal. At these angles the two bands coalesce and the reduction in conductivity is less than at neighbouring angles at which two separate bands are removed.

In Fig 8.6d there are three possible transitions in general, but for the practical case illustrated the largest energy difference (between the singlet and the lower level of the doublet) is neglected because at the appropriate temperatures there are few phonons of this energy. At a particular field B_0 the two energies are equal and only one phonon band is scattered instead of two at adjacent fields. At this field the reduction in conductivity shows a dip.

Fig. 8.7 shows examples which illustrate these ideally simple cases. Lanthanum cobalt nitrate contains cobalt ions which occur on inequivalent sites with different g_L values (thus corresponding to Fig. 8.6c) but at certain angles $g_L^X = g_L^Y$ and the level scheme is that illustrated in Fig. 8.6a.

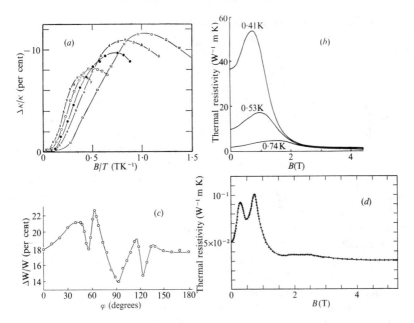

FIG. 8.7. Practical examples of the idealized cases illustrated in Fig. 8.6. (a) The fractional reduction in conductivity at various temperatures as a function of B/T for lanthanum cobalt nitrate: □ 1·10 K; △ 1·81 K; ● 2·38 K; ○ 3·27 K; × 4·32 K. The field was at 58° to the c-axis to reduce the two bands of Fig. 8.6(c) to one band as in Fig. 8.6(a). (After Brock and Huntley 1968.) (b) The thermal resistivity of praseodymium ethyl sulphate below 1 K as a function of magnetic field. At these temperatures the effective level scheme is as illustrated in Fig. 8.6(b). (After Harley et al. 1969.) (c) The fractional increase in thermal resistivity at 2·03 K and 1·33 T as a function of the angle θ between the field and the c-axis for lanthanum cobalt nitrate. (After Berman et al. 1963.) (d) The thermal resistivity at 4·25 K as a function of field (parallel to the c-axis) for holmium ethyl sulphate. (After McClintock et al. 1967.)

The fractional reduction in conductivity as a function of B/T is shown in Fig. 8.7a (Brock and Huntley 1968). On the simple model all the curves would coincide if $\tau(\omega, T)$ were represented by a simple function, but although this does not occur it can be seen that each curve passes through a maximum as predicted.

In praseodymium ethyl sulphate the ground state is split in the absence of a magnetic field. There is actually a higher level which must be considered if the temperature is not low enough, but the curves shown in Fig. 8.7b (Harley, McClintock, and Rosenberg 1969) are for temperatures low enough for this level not to be excited by the available phonons. For large enough fields the conductivity becomes larger than in zero field because the scattered band of phonons is far to the right in Fig. 8.6b. (In the figure, the resistivity is shown reaching a smaller value than in zero field.)

Fig. 8.7c shows the variation of thermal resistivity of lanthanum cobalt nitrate at constant temperature and field as the angle φ between the field direction and the crystal axis is varied (Berman et al. 1963). For $\varphi = 59°$, 121° the values of g_L^X and g_L^Y become equal and the dips at these angles occur because only one band of phonons is being scattered instead of two for neighbouring angles. If this band makes no contribution to the conductivity when it is scattered by one set of ions, the other set can affect the conductivity no further for this particular angle.

Fig. 8.7d shows the thermal resistivity of holmium ethyl sulphate at 4·25 K as a function of magnetic field (McClintock, Morton, Orbach, and Rosenberg 1967). At this temperature it is possible to neglect the transition with the largest energy difference, which is not indicated in Fig. 8.6d. Two bands of phonons are in general scattered, but for $B_0 = 0.55T$ these bands coalesce and there is a dip in the resistivity.

The effects which occur when the energy difference between different pairs of levels become equal and when levels cross have been studied in more detail by Anderson and Challis (1975). Many other systems have been examined, and measurement of thermal conductivity is now an illuminating adjunct to purely magnetic studies of crystals.

The model of a phonon system unaffected by the spins, except when an interaction takes place, is not an exact description of excitations in a paramagnetic crystal, but it is adequate for a low concentration of spins. More realistically, the spins and phonons form a coupled system in which the excitations are mixed (Jacobsen and Stevens 1963). Elliott and Parkinson (1967) considered the lifetime effects which result from this coupling and its effect on thermal conductivity. The general nature of the results is similar to that derived from the simple model, but the details are different.

It is possible for the spins to conduct heat themselves in an ordered magnetic system. The variations in the spin directions throughout the crystal are coupled, and the excitations occur as spin waves which have their own characteristic dispersion relation and have quantized energies, the quanta being called magnons (see, for example, Kittel 1971). Magnons can scatter phonons and also conduct heat themselves; Sato (1955) showed that on a simple model the thermal conductivity due to magnons would vary as T^2. The minimum magnon energies are raised by a magnetic field, so that fewer magnons are excited, and both the magnon conductivity and the magnon–phonon scattering decrease.

Friedberg and Douthett (1958) found a 20 per cent increase in the thermal conductivity of manganese ferrite at 2 K in a field of 1 T, which they attributed to a reduction in magnon–phonon scattering, while the magnons themselves made no appreciable contribution to the conductivity. Measurements on yttrium iron garnet have shown a reduction in a T^2

component of the conductivity with increasing magnetic field, ascribed to depopulation of the magnon states as their energy is increased (Lüthi 1962, Friedberg and Harris 1963, Douglass 1963). Metcalfe and Rosenberg (1972) made detailed measurements on gadolinium vanadate down to below 0·1 K and found sharp discontinuities in the conductivity occurring at 1 T, independent of temperature, which can be correlated with the appearance of the 'spin-flop' state between the antiferromagnetic and paramagnetic phases. The combination of magnons and phonons into a system of coupled modes has been treated by Walton, Rives, and Khalid (1973) and better agreement achieved with experiment than by addition of separate phonon and magnon conductivities.

8.3.3. Larger defects

Some work on aggregates of atoms will be described here, while dislocations are discussed separately in § 8.3.4.

There have been a number of experiments on the changes in conductivity produced by impurities precipitated to form clusters which have dimensions that are not small compared with all phonon wavelengths. Slack's KCl crystals containing Ca^{2+} (Slack 1957) were well annealed, and it was known that in supersaturated mixtures under these conditions platelets about 10 nm in diameter and 1·5 nm thick, which probably consist of $KCl.CaCl_2$, precipitate. At 5 K the dominant phonons in conducting heat in KCl have a wavelength ~7 nm, so that at some temperature above this the precipitates should scatter phonons nearly independently of their wavelength, while at some lower temperature the precipitates should be small enough compared with the phonon wavelengths for Rayleigh scattering to occur. On a simple model we would therefore expect that the conductivity over a temperature range below the position of the conductivity maximum should show a T^3 dependence, but with a mean free path less than the dimensions of the crystal. At lower temperatures the scattering by the precipitates should decrease, and the conductivity should be less temperature dependent until it has joined the T^3 curve corresponding to pure boundary scattering.

A behaviour as simple as this was found neither by Slack nor, in similar experiments, by Walker and Pohl (1963). The expected pattern was found by Worlock (1966) and Walton (1967) in measurements on sodium chloride containing silver colloids. Worlock's measurements extended down to 1·2 K and he observed a T^3 variation in conductivity, with the resistance which was additional to boundary scattering being proportional to colloid concentration. Walton (1967) extended the measurements on the same crystals down to 0·35 K, and observed that the conductivity then decreased more slowly than T^3 and at the lowest temperatures was returning to the values for the pure crystal. For a computed fit to the experimental results he used, for all polarizations, the scattering cross-section derived by Walton and Lee (1967) from the partial-wave analysis of Ying and Truell (1956) for longitudinal waves. This form of cross-section is similar to that given by Anderson (1950) and used by Schwartz and Walker (1967b). A colloid radius of ~6 nm gave the best fit, the quality of the fit being very sensitive to the radius assumed. From optical and microscopic studies Worlock had concluded that the particles had linear dimensions of the order of 20 nm or less, which is not inconsistent with a radius of ~6 nm.

Schwartz and Walker (1967b) made measurements on KCl and KBr crystals containing Ba^{2+} or Sr^{2+}, and, after quenching from 600 °C, the low-temperature conductivity, especially for crystals with KCl as host, changed from a T^3 dependence with a magnitude smaller than that of the pure crystal to another T^3 regime with a value equal to that of the pure host crystal, as can be seen in Fig. 8.8. They showed that the behaviour

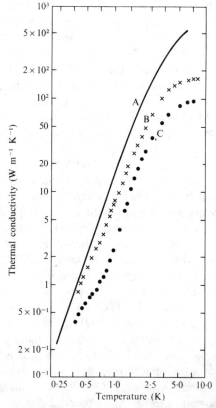

FIG. 8.8. Thermal conductivity of quenched KCl with divalent impurities: A, pure KCl; B, KCl with 5.8×10^{24} m^{-3} Ba^{2+}; C, KCl with 5.4×10^{24} m^{-3} Sr^{2+}. (After Schwartz and Walker 1967b.)

found can be represented qualitatively by a scattering rate which varies as ω^4 for low frequencies and is constant at the value it has for $qa = 1$ for frequencies above $\omega = v/a$, where a is the linear dimension of the precipitate. The centre of the transition region between the two types of conductivity behaviour occurs at a temperature which is inversely proportional to a. Again this temperature is close to that which would be given by the dominant-phonon method. The dominant phonons have the value $\omega = v/a$ in KCl when $T = 6/a$, where a is in nanometres. We would then expect that for 10 nm precipitates the centre of the change-over would occur at about 0·6 K. The more exact calculations, using both the simple representation of relaxation rates as divided into two regimes and a representation by Schwartz and Walker of Anderson's (1950) oscillatory function, give the centre of the cross-over region as ~0·7 K. For KCl containing both Ba^{2+} and Sr^{2+} the centre of the region was found to be between 0·5 and 0·6 K.

8.3.4. Dislocations

Experiments existed which showed that dislocations reduced the thermal conductivity of non-metallic crystals, but the first attempt to correlate the conductivity changes with the number of dislocations was made by Sproull, Moss, and Weinstock (1959). They compressed crystals of LiF to produce 2·6 and 4·0 per cent reductions in length and measured the etch-pit densities (1·8 and $4·6 \times 10^{11}$ m^{-2}) to give the density of dislocation lines intersecting the surface. Below the conductivity maximum the decrease in conductivity was so severe that to a good approximation boundary scattering could be neglected in comparison with the scattering by the dislocations. Between 2 and 8 K the conductivities of both deformed crystals were nearly proportional to T^2. The absolute values of the conductivity were compared with Klemens's predictions using eqn (8.3) with the appropriate constant, and were found to be over 10^3 lower. With this enormous discrepancy, the difference between the scattering by the strain fields of edge and screw dislocations can be neglected. Among the possible reasons for the discrepancy Sproull *et al.* considered but dismissed the possibility of scattering by mobile dislocations which had just been treated by Granato (1958). Moss (1965) later made measurements on CaF_2 as well as LiF and afterwards (Moss 1966) discussed these results together with those of Taylor, Albers, and Pohl (1965) on some other alkali halides. In all these cases, but not for germanium (Zaitlin and Anderson 1974a), the scattering was stronger than is given by either Klemens or Carruthers, and in NaCl and KCl the frequency dependence of the relaxation rate is not even clearly ω^1. It was suggested that the enhanced scattering arose from the association of dislocations.

Experiments which have been interpreted in terms of scattering by

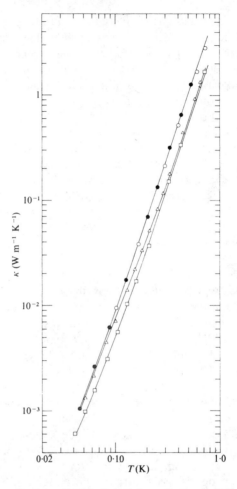

FIG. 8.9. Thermal conductivity of LiF: ● undeformed crystal; □ deformed crystal; △ 1000 R total irradiation; ○ 136 000 R total irradiation. (After Anderson and Malinowski 1972.)

vibrating dislocations were made on LiF both by Suzuki and Suzuki (1972) and by Anderson and Malinowski (1972). In the former experiments pinning of dislocations was produced after compression of the specimens by annealing at 300 °C for 10 min, while in the latter it was produced by irradiation with γ-rays after shearing the specimen. Anderson and Malinowski found that after a sufficient γ-ray dose the thermal conductivity of a deformed crystal recovers to the value it had for the crystal before deformation, as shown in Fig. 8.9. They concluded that after deformation the appreciable reduction in conductivity is due to scattering by mobile dislocations, and they used a combination of the models of Garber and Granato (1970) and of Ninomiya (1968). After γ-irradiation the dislocations would be rendered immobile by the point defects produced so that the dislocation scattering would then be by sessile dislocations, which was the case treated by Klemens and others. The upper limit to the scattering produced by such dislocations, as deduced from the experiments, is now so small that the previously calculated values could well be correct. In fact the scattering rates calculated by Ohashi (1968), which were several times larger than those of Klemens and Carruthers and at the time seemed to be an improvement as being nearer to the experimental values, now seem too high to account for the scattering by sessile dislocations.

9

AMORPHOUS SOLIDS

The solids which have been considered so far have been crystals which were either very nearly perfect or contained imperfections within such limits that there was still an underlying regularity of the lattice. There is, however, a large class of solids of practical importance in which there is no long-range order in the structure. Glasses belong to this category and so, to some extent, do many common polymers, a large number of which are of technological importance.

It was pointed out by Kittel (1949) that there is much less difference between the conductivities of amorphous solids than there is between the conductivities of crystals, and there have been suggestions that vitreous silica and nylon should be used as thermal-conductivity standards because the values found are little dependent on the particular sample measured.

There has been a renewed interest in the vibrational properties of amorphous solids, partly due to anomalous behaviour observed at very low temperatures; the specific heat does not follow the expected T^3 behaviour, and the conductivity does not follow a linear temperature dependence which was thought to correspond to the T^3 specific-heat variation. A review of recent theoretical and experimental work is given by Böttger (1974).

9.1. General behaviour of the conductivity

The conductivity of a crystal has been interpreted qualitatively in terms of the kinetic expression, in which the main features of the temperature dependence are given by the variations of the specific heat and of the mean free path. Above the conductivity maximum, the conductivity of a crystal decreases with increasing temperature because the mean free path decreases more rapidly than the specific heat increases. In an amorphous solid, however, the mean free path for short wavelengths is limited to the dimensions of the structural units, which are of the order of tenths of nanometres. At high temperatures, therefore, the conductivity follows the specific heat and decreases with decreasing temperature. Some substances can exist either in the crystalline or in the vitreous state, and at high temperatures the specific heats are not very different. For a crystal at high temperatures the phonon mean free path is also limited to the order of the lattice constant, so that the conductivities of the crystal and glass are not very different. As the temperature decreases the mean free path for the crystal increases much more rapidly than it does for the glass, and the ratio of the conductivities becomes very great. Below the temperature of the maximum, the mean free path in the crystal reaches a limiting value while at the same temperatures it still increases for the glass. Again the two conductivities come closer together. This is illustrated for crystalline and vitreous silica in Fig. 9.1, and the corresponding phonon mean free paths are shown in Fig. 9.2, with a few values tabulated in Table 9.1.

Carwile and Hoge (1966) considered the best values for the thermal conductivity of vitreous silica on the basis of over 20 sets of published measurements. From ~50 to ~500 K the measured conductivity followed the specific heat fairly closely, as is reflected in the small variation of phonon mean free path with temperature over this range, shown in Fig. 9.2. Above 500 K, however, the measured conductivities rose rapidly and by 1350 K some values were about three times greater than would be expected from the measured specific heat. Carwile and Hoge attributed

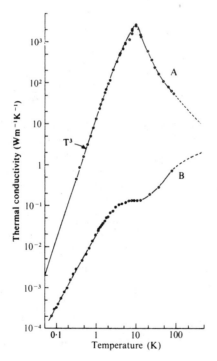

FIG. 9.1. Thermal conductivity of crystalline quartz parallel to the c-axis (rod with 5×5 mm^2 cross-section) (curve A) and of vitreous silica (curve B). (After Zeller and Pohl 1971.)

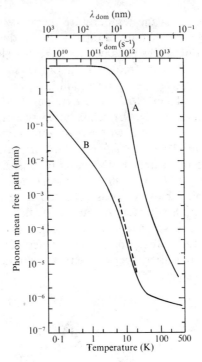

FIG. 9.2. Phonon mean free path in crystalline quartz (rod with 5×5 mm² cross-section) (curve A) and in vitreous silica (curve B). The broken line was computed on the mass-difference scattering model described in § 9.2. (After Zeller and Pohl 1971.)

the additional 'conductivity' to heat flowing through the glass by radiation, and showed that it could be explained quite well on a simple model of radiative transfer through a partially absorbing medium.

Estimates of the radiative component in 'conductivities' measured by steady-state methods are usually based on the analysis by Genzel (1953).

TABLE 9.1

Phonon mean free paths in crystalline and vitreous silica and their ratio at four different temperatures

Temperature (K)	l_{cryst} (mm)	l_{fused} (mm)	$l_{\text{cryst}}/l_{\text{fused}}$
500	2×10^{-6}	6×10^{-7}	3
10	7×10^{-1}	5×10^{-5}	14 000
1	5†	1×10^{-2}	500
0·1	5†	2×10^{-1}	25

† For a 5 mm diameter rod.

Men' and Sergeev (1973) analysed the effects in both steady-state and non-steady-state measurements. They measured the thermal diffusivity of vitreous silica and found that this also rose sharply above ~500 K, but when corrected for radiation the diffusivity continues to decrease gently, becoming constant above ~800 K, indicating a constant phonon mean free path. The true conductivities derived from the corrected diffusivity are in good agreement with those deduced by Carwile and Hoge from measured conductivities.

Below about 50 K the dominant wavelengths in vitreous silica become more than a nanometre and thus exceed the dimensions of the scale of the disorder in the structure. The mean free path consequently rises, and over a small temperature range just about keeps pace with the rate of fall in specific heat. The conductivity is nearly constant in the region of 10 K, before falling again at still lower temperatures. Below about 1 K the conductivity is proportional to T^w, where $w \sim 1 \cdot 8$.

Many non-crystalline solids have conductivities which behave in much the same way as that of vitreous silica, with a limiting low-temperature variation between $T^{1 \cdot 5}$ and T^2. Reese and Tucker (1965) measured Teflon, Kel-F, nylon, and polyethylene, and at the lowest temperatures the results show an approximately T^2 temperature dependence. For polymethyl methacrylate, polystyrene, and polyvinyl acetate, Choy, Salinger, and Chiang (1970) found a $T^{1 \cdot 5}$ dependence near 0·4 K and the same behaviour was found for polycarbonate (lexan) by Cieloszyk, Cruz, and Salinger (1973). Ashworth, Johnson, Hsiung, and Kreitman (1973) found no plateau in the conductivity curve for nylon; the conductivity increased a little faster than T above 1 K and was nearly constant above 100 K. Measurements on a particular specimen of polyethylene by Giles and Terry (1969) gave a conductivity varying faster than T^2 above 1 K and more slowly than $T^{1 \cdot 5}$ below this temperature.

Some 'non-crystalline' solids are made up of polymer chains which can arrange themselves in such a way as to form crystalline units, and it was not clear from early measurements whether the presence of some crystallinity in a specimen enhances or depresses the thermal conductivity. Choy and Greig (1975) have studied the thermal conductivity of polyethylene terephthalate in which the fraction of the volume occupied by crystalline material can be varied by suitable annealing after the original quenching. Fig. 9.3 shows the conductivity of seven such samples, in which the volume-fraction crystallinity, X, varied from effectively zero to just over $\frac{1}{2}$. Above ~20 K the more crystalline specimens have the larger conductivities, as might be expected, but below ~10 K the reverse occurs. This reversal in the effect of crystallinity probably explains the ambiguity of earlier results.

Choy and Greig explained their observations as the consequence of the

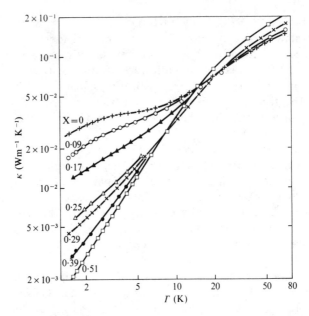

FIG. 9.3. The thermal conductivity of polyethylene terephthalate with different values of the volume-fraction crystallinity X. (After Choy and Greig 1975.)

competing influences of the higher conductivity of the crystalline regions and of the thermal resistance which arises at the junctions between amorphous and crystalline regions. They used a simple model of the arrangements of these different regions, and applied Little's (1959) theory for the resistance at the interface between materials with different elastic properties to compute the overall conductivity. This theory shows that the contact resistance is a strong function of temperature ($\propto T^3$ at low temperatures for non-metals in complete mechanical contact), and at low temperatures the resistance at the many interfaces far outweighs the lower resistance of the crystalline regions themselves. As boundary resistance becomes more important, the conductivity curves become steeper. Choy and Greig obtained quite good agreement between the conductivities computed on their model and the measured values.

Contact resistance also explains the effects of introducing crystalline materials of relatively high conductivity into epoxy resins (Garrett and Rosenberg 1972). At high temperatures the conductivity is increased by the presence of a high-conductivity filler, but below a certain temperature it is decreased by the presence of internal boundaries which give rise to a resistance which increases with decreasing temperature.

We may consider the border between the high- and low-temperature behaviours of the conductivity of a non-crystalline solid, such as vitreous silica, to be around liquid-helium temperatures. Above these temperatures the conductivity accords with the idea that with falling temperature the phonon mean free path should increase as the dominant wavelengths become much larger than the scale of the disorder in the structure. Below liquid-helium temperatures, however, the specific heat is larger than that predicted by the usual theory for a solid with the measured elastic constants. If additional excitations contribute to the specific heat, they may not contribute equally to the thermal conductivity, so that a mean free path derived from a comparison of specific heat and conductivity would be less meaningful than it is for a crystal.

9.2. The 'high'-temperature behaviour

Kittel (1949) first explained the thermal conductivity of glass in terms of the corresponding mean free path. He pointed out that at normal temperatures this was constant at a value of a few tenths of a nanometre, which is about the size of the SiO_4 tetrahedra which form the irregularly repeating units in vitreous silica. He showed that the mean free path starts to increase when the dominant phonon wavelengths become larger than this unit, and the plateau observed soon afterwards around 10 K by Berman (1949) reflects a smooth increase in mean free path as the temperature decreases.

Klemens (1951) developed a more elaborate theory to explain the temperature variation of the conductivity in terms of the separate contributions of longitudinal and transverse phonons. The atomic vibrations were still considered to be resolvable into normal modes, but these are not plane waves and there is interaction between them, which Klemens called 'structure scattering'. The conductivity was taken to be the sum of contributions from the different polarizations and was given in terms of three adjustable constants. When these were fitted to give the best agreement with measurements on vitreous silica, it was found that at high temperatures κ_{trans} predominates and is proportional to T, while κ_{long} decreases slightly with increasing temperature. At low temperatures κ_{long} is about 40 times κ_{trans}, both being proportional to T. At ~10 K κ_{long} has a fairly flat maximum, and the observed plateau around 10 K marks the transition from κ_{long} dominant and nearly constant to κ_{trans} dominant and proportional to T at higher temperatures.

Ziman (1960) discussed the analogy between the propagation of phonons in a glass and the propagation of radio waves in an irregularly refracting ionosphere. Again the relaxation rate for short waves was found to be constant and for long waves to be proportional to q^2, giving a conductivity proportional to T at low temperatures.

As another model Klemens (1965) suggested that the heat-carrying phonons in a glass could be scattered by localized phonons in a resonant way which would yield the plateau around 10 K in a similar way to the formation of the dips found for crystals containing certain substitutional molecules (see § 8.3.2(a)). From measurements of the elastic constants the specific heat of a glass at low temperatures would, according to the Debye theory, be slightly greater than that of the corresponding crystal. However, while the measured specific heat of crystalline quartz at low temperatures tends to the value predicted from the elastic constants, the specific heat of the glass stays considerably higher than that calculated (Flubacher, Leadbetter, Morrison, and Stoicheff 1959); a similar discrepancy was later found for polymethyl methacrylate and polystyrene by Choy, Hunt, and Salinger (1970). Dreyfus, Fernandes, and Maynard (1968) suggested that the extra modes which give the increased specific heat might be the localized modes producing the resonance scattering.

Zeller and Pohl (1971) did not consider that any of these scattering theories could explain why so many non-crystalline materials had such similar conductivities, and gave an order-of-magnitude calculation which led to the similarities observed. They assumed that every atom is displaced from a regular array and that the scattering source is represented by a vacancy at every lattice site. The scattering is easily calculated since $\Delta M/M = 1$. The effective volume was taken as the size of the appropriate atom (for Se) or molecule (for SiO_2 and GeO_2). The mean free paths were calculated for the dominant phonons as a function of temperature, and in the vicinity of 10 K the agreement with those deduced from the conductivity is remarkably close, as can be seen in Fig. 9.2. Zeller and Pohl did not claim too much significance for this calculation, but it has the merit of being a simple basis for expecting similar mean free paths and conductivities for a wide range of non-crystalline solids.

9.3. The conductivity at very low temperatures

Measurements which extended down to only 1 or 2 K suggested that the limiting temperature variation of the conductivity of several amorphous solids would be as T^1, in agreement with Klemens's theory. However, there were results which indicated a more rapid temperature dependence, and these were reinforced by measurements made down to below 0·1 K on many non-crystalline solids (see, for example, Stephens 1973) which showed that the conductivity, in fact, seems to vary only slightly more slowly than T^2. Specific-heat measurements have also been extended down to these low temperatures and the anomaly already mentioned has been found to be of the same nature for all these amorphous solids. Not only is the specific heat C_{exp} larger than would be expected from the Debye theory using the measured elastic constants, but its temperature variation is much slower than that of the Debye specific heat C_D, so that the ratio C_{exp}/C_D increases rapidly as the temperature is decreased below

1 or 2 K. The experimental data can be expressed in the form

$$C_{\exp} = C_1 T + C_2 T^2 + C_3 T^3 + \ldots$$

corresponding to a density of states for excitations

$$f(E) = f_1 + f_2 E + f_3 E^2 + \ldots$$

and there is little to choose between an expression of the form

$$C_v = C_1 T + C_3 T^3$$

and polynomials containing more terms (Stephens 1973). With this representation, C_3 is up to nearly three times larger than the value derived from the elastic constants on the Debye model, and the linear term is dominant below $\sim 0{\cdot}2$ K. For vitreous silica the two terms are equal at $\sim 0{\cdot}7$ K. It was noted by Zeller and Pohl (1971) that the linear term does not vary greatly between different non-crystalline solids and is not dependent on the purity of the particular specimen. Later experiments (Stephens 1976) show that the specific heat can be appreciably affected by impurities, although the thermal conductivity is practically unchanged.

The nature of the additional excitations has been the subject of much study, but the problem is by no means resolved (see, for example, Leadbetter 1972). A large cubic component in C_V could come from an enhancement of the quadratic part of the phonon density of states for low ω. It is more illuminating to compare the properties of vitreous silica with those of crystobalite, rather than quartz, since their densities and short-range structures are more alike. Bilir and Phillips (1975) measured the specific heat of crystobalite down to 2 K and found that between 12 and 15 K it is about five times that given by the Debye theory. They explain this in terms of a low-lying, rather flat transverse acoustic branch of the dispersion curves, as was suggested by Leadbetter (1968). The behaviour of the specific heat of vitreous silica above temperatures at which the linear term is important is, in fact, not very different from that of crystobalite; the equivalent peak in $C_{\exp}/C_D$ is lower and occurs at a lower temperature (~ 10 K) than for crystobalite. Leadbetter (1969) carried out inelastic cold-neutron scattering experiments on both materials and found a sharp peak in $f(\omega)$ for crystobalite at 40 cm^{-1}, which was much broadened in vitreous silica. Phonons of this frequency would be dominant in their contribution to the heat capacity at ~ 15 K.

Several models have been proposed to account for a linear component in the specific heat, but most of the recent comparisons between experiment and prediction have been made on the basis of the model of Anderson, Halperin, and Varma (1972). They suggested that atoms or groups of atoms tunnel between equilibrium positions at almost the same energy, with a smooth distribution over energy difference of the number of levels between which tunnelling is possible. This model would apply to disordered systems in general and can be made to yield the linear contribution to the specific heat. The T^2 variation in conductivity then results from scattering of phonons by these tunnelling modes.

The theory predicted that the ultrasonic attenuation should increase at low temperatures, and this was observed by Hunklinger, Arnold, and Stein (1973). In addition, an observed increase in attenuation with decreasing power can be explained by the model (Hunklinger, Arnold, Stein, Nava, and Dransfeld 1972, Golding, Graebner, Halperin, and Schutz 1973), as can the initial increase in ultrasonic velocity with increasing temperature above 0·28 K (Piché, Maynard, Hunklinger, and Jäckle 1974). The very large value of the Grüneisen parameter γ (-40 to -50), deduced from the expansion coefficient of vitreous silica at very low temperatures (White 1975), is another observation which has been held to be explicable by the model.

A similar model was proposed at about the same time by Phillips (1972), who explained the much larger phonon relaxation time deduced from Brillouin scattering than that deduced from thermal conductivity by the saturation of the resonance transitions by the incident radiation.

Morgan and Smith (1974) considered the elastic scattering of phonons by fluctuations in the relevant properties from their mean values. They showed that the conductivities below 1 K could be explained by using a long correlation length (100 to 300 nm) for these fluctuations. The similarity between the conductivities of different non-crystalline solids then arises from the fact that a range of combinations of the two parameters in the theory—the correlation length and the magnitude of the fluctuations—can produce a similar behaviour below 1 K.

Zaitlin and Anderson (1974b) showed that, if the bulk scattering above helium temperatures is due to both the tunnelling mechanism and 'point-defect' scattering (Zeller and Pohl 1971, mentioned in the previous section), then a plateau should occur in the conductivity at a temperature little dependent on the particular amorphous material. This results from the temperature of the plateau being proportional to the cube root of the ratio of the constants which describe the strengths of the two scattering mechanisms.

Love (1973) observed Brillouin scattering in two glasses which indicated the existence of transverse and longitudinal phonons with velocities which agreed well with the values derived from the elastic constants. In addition, the intensities of the scattering decreased linearly with temperature, which is as expected for phonons. The lifetime of the phonons could be estimated from the width of the Brillouin lines and was more than 10 cycles at all temperatures. It therefore appears reasonable to conclude that part of the T^3 term in the specific heat arises from phonons with similar properties to those in crystals

An indication that thermal conductivity is mainly attributable to 'ordinary' phonons comes from experiments by Stephens (see Pohl, Love, and Stephens 1974) and by Zaitlin and Anderson (1974b, 1975). Stephens measured the thermal conductivity of thin soda-silica glass fibres of 60 μm diameter. At 0·1 K the phonon mean free path in bulk vitreous silica is $\sim 2 \times 10^{-4}$ m (the exact value depends on the value taken for the heat capacity of the carriers), so that boundary scattering would be expected to show up in the fibres, which have diameters about $\frac{1}{3}$ of this mean free path. The surfaces of the fibres as drawn were too smooth to produce appreciable boundary scattering, but for etched fibres the conductivity was, indeed, less than for bulk glass. The conductivity of the fibres could

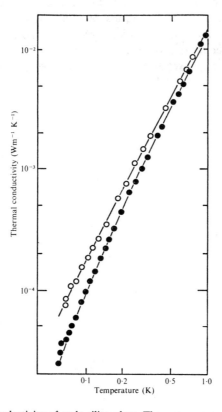

FIG. 9.4. Thermal conductivity of soda-silica glass. The upper curve was measured on a bulk specimen and the upper points are for smooth fibres. The lower curve was calculated for 60 μm diameter rough fibres on the assumption that only 'Debye phonons' take part in the conduction. The lower points were measured on rough fibres of 60 μm diameter. (After Pohl et al. 1974.)

be fitted by adding a boundary resistance to the thermal resistivity of the bulk material, if it was assumed that all the heat is conducted by phonons which contribute a specific heat as given by the Debye theory and computed from the elastic constants. With this value for the specific heat and with a mean phonon velocity also computed from the elastic constants, the boundary-scattering mean free path deduced from the experiments was just 60 μm. The good agreement between the measured and computed conductivities of the fibres is shown in Fig. 9.4.

Zaitlin and Anderson (1974b) came to the same conclusion from measurements on a borosilicate glass in which 'boundary' scattering was caused by the presence of a large concentration of small holes. They have confirmed this more recently (Zaitlin and Anderson 1975) and also offered an explanation of the plateau at ~ 10 K in terms of a rapid decrease in phonon mean free path with increasing frequency.

On this picture of Debye phonons carrying the heat, the much smaller sensitivity of the conductivity, compared with the specific heat, to impurities can be accounted for by a weak coupling between these phonons and those low-energy excitations which make the specific heat sensitive to impurities (Hunklinger, Piché, Lasjaunias, and Dransfeld 1975).

10
ELECTRONS

THE thermal conductivities discussed so far have been due mainly to transport of energy by lattice vibrational modes. Heat transfer by radiation and heat conduction by magnons have been mentioned briefly, but otherwise no other mechanism was invoked. In solids which are usually thought of as good heat conductors, however, heat is mainly transported by electrons, and, although a few exceptional non-metals (see § 7.1.1(b)) have high thermal conductivities at normal temperatures, the most common good heat conductors are metals. In metals such as copper and silver the electronic thermal conductivity is so dominant that to a very good approximation the observed conductivity at all temperatures up to the melting point can be ascribed entirely to electrons. In other metals, such as antimony and bismuth, and in many alloys the lattice conductivity is comparable with that due to the electrons, and over some temperature ranges even exceeds it.

In superconductors the electronic thermal conductivity decreases below the superconducting transition temperature and the lattice conductivity gradually becomes dominant. As the transition between the normal and superconducting states of some metals can be brought about at low temperatures by small magnetic fields, a simple heat switch can be based on the difference between almost purely electronic thermal conductivity and almost purely lattice thermal conductivity.

In semiconductors the relative importance of the two mechanisms of heat transport is determined by the doping of the material.

An outline of the behaviour of electrons in metals is given in this chapter, and their scattering is considered in Chapter 11. The consequences for thermal conductivity in metals and alloys are discussed in Chapter 12 and in semiconductors in Chapter 13.

There is much in common between the general methods of treating conduction of heat when electrons are the carriers and those used earlier for discussing heat transport by phonons. There is a very close relationship between the calculations of relaxation times appropriate to thermal conductivity and to electrical conductivity, evident in the Wiedemann–Franz–Lorenz law. In order to avoid too much repetition of what has already been said here on phonons and in other books on electrical conductivity of metals, alloys, and semiconductors (e.g. Blatt 1968), the review of the theory will be considerably shortened.

10.1. Drude theory of conductivity

The concept of electrons as particles is easier to grasp than the concept of phonons, and, although one must be aware of the wave aspects for a proper understanding of the motion of electrons, theories of electrical and thermal conductivities of metals capable of explaining the gross features date from the beginning of the century. Such theories can explain qualitatively the Wiedemann–Franz law (1853) and its extension by Lorenz (1881), both of which were first found experimentally.

In the Drude theory (1900, 1902) it was assumed that the electrons in a metal behave as independent gas-like particles with random thermal velocities. In the absence of an electric field or temperature gradient there are equal numbers of electrons with opposite velocities for all directions, and the net flow of charge and energy is zero. In an electric field $\mathscr{E}$ the electrons are accelerated in one direction, so that a drift velocity is superimposed on the random motion and there is a net flow in the direction of the field (against the field for negatively charged electrons). If the drift velocity of an electron is destroyed by some sort of interaction with the lattice, in such a way that the probability of an electron having escaped a 'collision' after a time t is $\exp(-t/\tau)$, the mean drift velocity is $(\mathscr{E}e/m_e)\tau$, where e and m_e are the charge and mass of an electron. If there are n_e electrons per unit volume, the resulting current density $\mathscr{J} = (n_e e^2 \tau \mathscr{E})/m_e$ and the electrical conductivity $\sigma = (n_e e^2 \tau)/m_e$. The expression for the thermal conductivity is similar to that derived for phonons: $\kappa = (n_e c_e v^2 \tau)/3$, where c_e is the heat capacity per electron ($n_e c_e = C_e$, the heat capacity per unit volume contributed by the electrons), and v is their mean thermal velocity (which is much greater than the drift velocity). The Wiedemann–Franz–Lorenz (WFL) law is obtained by finding the ratio of κ to σ and assuming that the same relaxation time τ applies for consideration of electrical and thermal conductivity, so that the separate τ's cancel out in the ratio

$$\frac{\kappa}{\sigma} = \frac{1}{3} \frac{c_e v^2 m_e}{e^2}. \tag{10.1}$$

Treating electrons as gas-like particles which obey the law of equipartition of energy, $c_e = \tfrac{3}{2} k_B$ and $\tfrac{1}{2} m_e v^2 = \tfrac{3}{2} k_B T$. Eqn (10.1) then becomes

$$\frac{\kappa}{\sigma} = \tfrac{3}{2} T \left(\frac{k_B}{e}\right)^2 \tag{10.2}$$

which gives the relation $\kappa/\sigma T = \text{constant}$.

ELECTRONS

We shall see in the next section that a proper treatment of free electrons leads to a specific heat proportional to temperature and a velocity independent of temperature, so that with the same assumption of equal relaxation times the ratio $\kappa/\sigma T$ is still constant but the numerical factor becomes $\pi^2/3$ instead of $\frac{3}{2}$. This assumption of equal relaxation times implies that the non-equilibrium situations produced by an electric field and by a temperature gradient relax to the thermal equilibrium state at the same rate. This assumption is, however, only valid in certain circumstances, as discussed in § 10.4.

Even in this rather crude explanation of electrical and thermal conductivities due to electrons, one important difference from the phonon case has been taken for granted. In pure non-metals it was always assumed that the conductivity is mainly determined at normal temperatures by the scattering of phonons by one another (however this notion may have been expressed at the time of the relevant theory). In metals it has generally been assumed that the scattering of electrons by one another can be ignored (this is, in fact, largely justified; see § 11.3.2). It is only at low temperatures that there is some similarity between the scattering processes, in that for both phonons and electrons the conductivity is then determined by 'defects' in the lattice. Even so, there is a difference, since for reasonably pure specimens phonon conduction in non-metals at low temperatures is limited by the finite extent of the lattice, whereas the electronic conductivity in metals is limited by more commonplace lattice defects.

10.2. Electrons in metals

10.2.1. Free electrons

Although electrons in a metal are not at all free, some of their essential properties required for understanding conduction can be derived from considering them as a gas of weakly interacting particles moving about inside an empty box. As for lattice vibrations, it is easiest to think first of one-dimensional motion. If there is no variation in potential over the length of the box, then if as in § 4.1.1 we assume periodic boundary conditions the wavelength of the electronic wave function must be such that an integral number of waves just fit the length L. The possible wavelengths in descending order are thus $L, \frac{1}{2}, \frac{1}{3}L, \ldots$ and the corresponding wave-numbers k are $2\pi/L, 4\pi/L, 6\pi/L, \ldots$. In three dimensions each component of the wave-vector k can only take one of the values $\pm 2\pi/L$, $\pm 4\pi/L, \ldots$ if the box is a cube of side L (the shape of the box actually has no effect on the macroscopic properties of the electrons unless their density is far smaller than occurs in practice). The momentum of an electron is $\hbar k$, so that for a free electron the possible energies are

$$E(k) = \frac{(\hbar k)^2}{2m_e}$$

$$= \frac{\hbar^2}{2m_e}(k_x^2 + k_y^2 + k_z^2)$$

$$= \frac{2\pi^2 \hbar^2}{L^2 m_e}(n_x^2 + n_y^2 + n_z^2) \qquad (10.3)$$

where n_x, n_y, and n_z are the integral quantum numbers describing the energy of an electron. Associated with their half-integral spin, electrons obey the Pauli exclusion principle, and only two electrons may have the same set of quantum numbers n_x, n_y, n_z and these two must then have opposite spins. The lowest energy for a collection of electrons is thus greater than that corresponding to each electron having the minimum energy it could have if it were alone in the box. Each energy level defined by eqn (10.3) can only be occupied by two electrons, so that if there are N electrons to be accommodated they will occupy the first $N/2$ levels. If we are dealing with the sort of electron densities which occur in metals, we need to go up to very high values of the n's, and if we imagine a three-dimensional plot with axes k_x, k_y, and k_z of the combinations of k_x, k_y, and k_z all of which give a particular value of this high energy, they will be represented to a very good approximation by points on the surface of a sphere of radius k_{max}. Compared with the unit of $2\pi/L$, k_{max} is so large that the discreteness of the lattice of points on this plot is not noticeable. This is illustrated in two dimensions in Fig. 10.1.

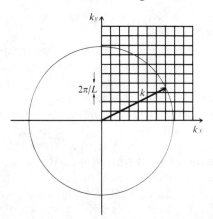

FIG. 10.1. Possible values for the components k_x, k_y, and k_z of the electron wave-vector are multiples of $2\pi/L$. A sphere of radius k and volume $\tfrac{4}{3}\pi k^3$ contains $\tfrac{4}{3}\pi k^3/(2\pi/L)^3$ combinations of components, represented by the grid intersections, if the scale of the grid (shown for positive k_x and k_y) is fine enough compared with k.

We may consider that unit volume of three-dimensional k-space contains $(L/2\pi)^3$ levels (many of which have the same energies), so that we need to go up to a volume $(N/2)(2\pi/L)^3$ in k-space to accommodate N electrons. The maximum value of the corresponding energy is called the Fermi energy E_F, and the maximum wave-number is written k_F. We have

$$\tfrac{4}{3}\pi k_F^3 = \frac{N}{2}\left(\frac{2\pi}{L}\right)^3$$

so that

$$k_F = \left(\frac{3N}{8\pi}\right)^{\tfrac{1}{3}} \frac{2\pi}{L} = \left(\frac{3\pi^2 N}{V}\right)^{\tfrac{1}{3}} = (3\pi^2 n_e)^{\tfrac{1}{3}} \qquad (10.4)$$

writing the volume V for L^3. The Fermi energy is

$$E_F = \frac{(\hbar k_F)^2}{2m_e} = \frac{\hbar^2}{2m_e}(3\pi^2 n_e)^{\tfrac{2}{3}} \qquad (10.5)$$

and the velocity of the electrons with this energy is

$$v_F = \frac{\hbar k_F}{m_e} = \frac{\hbar}{m_e}(3\pi^2 n_e)^{\tfrac{1}{3}}. \qquad (10.6)$$

These values of k_F, E_F, and v_F characterize the most energetic electrons at absolute zero, when the electron system has the lowest energy consistent with the exclusion principle. For copper E_F is 7 eV and v_F is $1 \cdot 6 \times 10^6$ m s^{-1}. The mean energy per electron is $\tfrac{3}{5}E_F$ and the mean velocity is $\tfrac{3}{4}v_F$, so that at 0 K the average electron velocity in copper is $\sim 10^6$ m s^{-1}, which is rather different from the value zero given by classical theory. On classical theory the average velocity even at room temperature would be only 7×10^4 m s^{-1}. Similarly the average energy per electron at 0 K is ~ 4 eV, while if electrons behaved as classical particles the average energy at room temperature $(\tfrac{3}{2}k_B T)$ would be only 4×10^{-2} eV.

If the electrons are not at 0 K, some of the energy levels below E_F are empty and an equal number above E_F are occupied. Electrons obey Fermi–Dirac statistics and the average population of a state with energy E_i and a particular spin direction is given by the Fermi–Dirac distribution

$$\mathscr{F}(E_i) = \frac{1}{\exp\{(E_i - E'_F)/k_B T\} + 1} \qquad (10.7)$$

where the reference energy has been written as E'_F to indicate that at a finite temperature its value is slightly different from the Fermi energy calculated above for 0 K. In fact, E'_F is the chemical potential.

For $T = 0$ K, $\mathscr{F}(E_i)$ is 1 for $E_i < E_F$ and 0 for $E_i > E_F$. At any finite temperature $\mathscr{F}(E_i) \to 1$ as $E_i \to 0$, and is in fact 0·98 for E_i only $4 k_B T$ below E'_F. For E_i the same amount above E'_F, $\mathscr{F}(E_i)$ has fallen to 0·02. This means that for copper at room temperature $\mathscr{F}(E_i)$ falls from 0·98 to 0·02 over an energy interval which is ~3 per cent of E'_F.

The heat capacity of a free-electron gas of N particles may be derived from the results given and is found to be

$$C_e = \tfrac{1}{2}\pi^2 N k_B \frac{k_B T}{E_F} \tag{10.8}$$

for temperatures such that $k_B T \ll E_F$ (for ordinary metals this merely limits the temperature to much less than 10^4–10^5 K). We would arrive at an expression for the heat capacity similar to eqn (10.8) by considering that only a fraction $\sim k_B T/E_F$ of the N electrons can contribute $\tfrac{3}{2} k_B$ to the heat capacity, since only those with energies near E_F have unoccupied energy levels to go to when they take up more energy.

Although the Drude theory of conduction can be improved by treating the electrons as free and yet as being scattered by the lattice, the existence of a lattice in which the electrons move has profound effects on the electron states themselves, even in the absence of applied fields.

10.2.2. *Electrons in a crystal lattice*

In a crystal lattice the potential seen by the electrons varies regularly with position, and the wave-functions become the product of a plane-wave solution, applicable to free electrons, and a function which has the periodicity of the lattice—a Bloch function. The waves still propagate without attenuation in a perfectly regular lattice. The presence of the lattice affects the variation of the energy of an electron with wave-number (for free electrons the relation is quadratic) and the possible energies for propagating electrons to have in a lattice. If we consider a simple cubic lattice, as in § 4.4.1 for phonons, then for an electron wave vector of such a magnitude and direction that it reaches nearly to the perimeter of the Brillouin zone the energy is appreciably different from the energy for the same value of k on a free-electron model. For $|k| < \pi/a$ the energy is less than the free-electron value and for $|k| > \pi/a$ it is greater. This means that there is an energy gap at the zone boundary and there is no propagating solution of the wave equation which gives energies lying within this gap. For low k the E–k relation becomes the same as for free electrons, as is illustrated in Fig. 10.2 in one dimension. It can be understood that something will occur for values of k lying close to the zone boundary, since the condition for Bragg reflection of waves is that the wavelets emanating from successive rows of atoms should be in phase. In one dimension this means that the distance between two atoms should

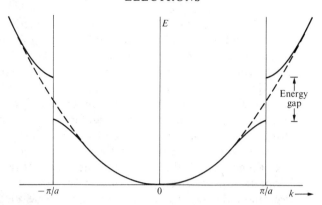

FIG. 10.2. One-dimensional representation of the energy of an electron as a function of wave-number for a periodic potential. The broken line shows the relation for free electrons. (After Ziman 1963.)

be one half-wavelength, so that $a = \frac{1}{2}\lambda = \pi/k$, or $k = \pi/a$, which is the perpendicular distance from the origin to the side of the Brillouin zone. The same principle applies in three dimensions and the sides of the cubic zone delineate the values of k at which the energy gap occurs for electrons in a simple cubic lattice. The waves represented by these values of k are stationary waves (the repeated Bragg reflections prevent them from travelling) and their group velocity, given by the slope of the E–k curve, is zero. Just as for phonons and atomic vibrations, so $k = \pi/a$ represents the same electronic state as $k = -\pi/a$, and when we count up the number of electron energy states for k inside the first Brillouin zone we find exactly N distinct states, where N is the number of atoms. Since each state can contain two electrons of opposite spin, $2N$ electrons can be accommodated within this range of k.

10.2.3. Metals, insulators, and semiconductors

The first Brillouin zone can accommodate $2N$ electrons, so that if the Fermi surface for a metal with one electron per atom were still a sphere even in a crystal lattice, its volume would be half the volume of the cubic zone. The radius of this sphere, r, would be such that $\frac{4}{3}\pi r^3 = \frac{1}{2}g^3$, where g is the reciprocal lattice vector, so that $r = 0.98(g/2)$ and the occupied k values would extend to within 2 per cent of the zone edge in the directions of the perpendiculars from the zone centre to the faces. Such values of k are so near to those at which the energy gap appears in these directions that the actual energy they represent may be considerably below the value expected from the free-electron model. On the other hand, the same values of k in the directions of the diagonals of the zone are only a little over half-way to the corners of the cubic zone and the

E–k relation may be little different from the free-electron relation. Electrons will tend to fill up levels with greater k values in the directions of the perpendiculars to the faces, since they can thereby have lower energies than by continuing to fill up states along the diagonals in k-space. This tendency is illustrated in Fig. 10.3, which shows that the Fermi surface in a real metal can be very different from that for a system of free electrons (in a two-dimensional representation the radius of the circle representing the Fermi surface for free electrons, equal to 0·98 g/2, appears to be 70 per cent of the distance to the corner of the square, rather than under 60 per cent of the distance to the corner of the cube).

If the solid contains one electron per atom (if we are to consider any crystal structure, then the equivalent requirement is one electron per primitive cell), the Fermi surface can lie entirely within the first Brillouin zone, leaving half its volume empty. For any arbitrary energy gaps in different regions of the zone boundary the occupied volume in the first zone can never be more than half, so that there will always be states available to electrons at the Fermi surface with energies immediately adjacent to their own. Electrical conduction entails accelerating electrons in an electric field, thereby taking them to states of different energy. A crystal with one electron per atom must therefore be a metal.

A crystal with two electrons per atom could have all its electrons exactly occupying the first zone. Because of the energy gap at the zone boundary, none of the electrons has a state with energy close to its own which is not already occupied, and such a crystal would be an insulator. However, if the energy gap across the zone-boundary faces in the directions of the perpendicular bisectors is not too great, the energy of the states just across this boundary may be less than those corresponding to values of k directed along the diagonals. Some of the electron levels

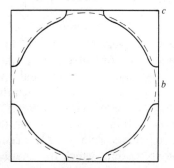

FIG. 10.3. Two-dimensional representation of the Fermi surface in a periodic potential. The energy of electrons with wave-vectors represented by points b can be less than the energy for the same wave-number but with the vector directed towards the corners c. States near b are occupied up to the limit of the Brillouin Zone and the Fermi surface differs from the free-electron sphere shown by the dotted circle.

with k in the diagonal directions may therefore be unpopulated, while some levels in the next zone may be populated. Electrons at the Fermi surface then have vacant adjacent levels and can be accelerated by fields. It is for this reason that the alkali earth metals with two electrons per atom are metals.

Crystals with an odd number of electrons per atom must be metals, while those with an even number can be insulators or metals, depending on the energetically favourable way of fitting the Fermi surface to the Brillouin zone. There is a category of semi-metal which covers cases where there are partially filled zones, but the number of empty states or filled states in higher zones is very small.

In an insulator the electrons just fill a whole number of zones, and the energy gap at the zone boundary is so large that at ordinary temperatures the probability of an electron being thermally excited across the gap is negligibly small. However, if the gap is small, there will be some electrons excited from the full band into the empty band, and conduction is possible. As the electrons which conduct are thermally excited, their number depends on temperature, which is a characteristic of a semiconductor. In many semiconductors of practical importance the electrons are produced by suitable doping of the crystal. Empty levels can be introduced just above a filled band into which electrons can be thermally excited because of the small energy gap, or electrons can be introduced with energies just below an empty band and can be thermally excited into it. In both cases the number of empty states in the lower band or filled states in the upper band will be dependent on temperature, and the conductivity of semiconductors should be zero at 0 K.

10.3. Conduction by electrons

In this section it will be assumed that there are processes which tend to restore the thermal equilibrium electron distribution given by eqn (10.7), some being an intrinsic property of the material and others arising from imperfections in the particular specimen. From a simple treatment of nearly free electrons we can see the conditions under which the WFL law should hold and how departures from it can arise. It must be emphasized that we are only considering the relation between electronic heat conduction and electrical conductivity. Any apparent departures from the law because the measured thermal conductivity also includes a lattice component are irrelevant to this discussion.

In an electric field $\mathscr{E}$ electrons are accelerated, and the analogue of the classical expression for the rate of change of momentum is $\dot{\mathbf{k}} = (e/\hbar)\mathscr{E}$ ($\hbar k$ is analogous to momentum). This relation still holds for electrons in a periodic lattice. After a time t the values of k for all electrons have increased by $(et/\hbar)\mathscr{E}$. If on the average an electron at the Fermi surface

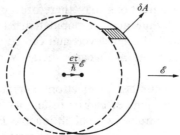

FIG. 10.4. The displacement of the Fermi surface in an electric field $\mathscr{E}$; δA is an element of area in k-space on the Fermi surface. The shaded area $\delta A(e\tau/\hbar)\mathscr{E}$ represents $(2/8\pi^3)\delta A(e\tau/\hbar)\mathscr{E}$ electrons per unit volume. (After Ziman 1963.)

accelerates for the relaxation time τ, a stationary state will be reached in which the whole sphere representing the occupied electron states has shifted by an amount $(e\tau/\hbar)\mathscr{E}$ in the direction of the field, as shown in the k-space representation of Fig. 10.4 (for simplicity in drawing, it is assumed that the Fermi surface is a sphere). In this condition the rate at which electrons are 'picked off' the Fermi surface at the right and are led round to the left by scattering processes just matches the rate at which electrons are brought up to the position of the right-hand spherical surface from levels just below it. The current density which results from this shift in the Fermi surface is given by the product of the velocity of the electrons, their number per unit real volume, and their charge. The number per unit real volume represented by the shading in the volume swept out by the area $\delta \mathbf{A}$ in moving δk to the right is $(2/8\pi^3)\delta \mathbf{A} \cdot (e\tau/\hbar)\mathscr{E}$ and each electron has the Fermi velocity in a direction parallel to $\delta \mathbf{A}$. The total current density is

$$\mathscr{J} = \frac{1}{3}\frac{e^2\tau}{4\pi^3\hbar}v_F A_F \mathscr{E},$$

where v_F, by eqn (10.6) is $(\hbar/m_e)(3\pi^2 n_e)^{\frac{1}{3}}$ and A_F by eqn (10.4) is $4\pi(3\pi^2 n_e)^{\frac{2}{3}}$. The conductivity is thus

$$\sigma = \frac{n_e e^2 \tau}{m_e}. \tag{10.9}$$

This is just the same as the classical expression (§ 10.1) derived on the assumption of an exponential distribution of times for acceleration before collisions. The even simpler classical derivation, on the assumption that every electron accelerates for exactly the same time before its drift velocity is reduced to zero, could be represented by the sphere in Fig. 10.4 moving to the right for a time τ and then flicking back to its initial position before again being accelerated to the right, the average displacement being $\delta k/2$, so that expression (3.1) appears to be half that given in eqn (10.9).

ELECTRONS

The same value for the electrical conductivity is obtained if we start with the Boltzmann equation, as was done for phonons in § 4.2.1; τ is then a measure of the rate at which a non-equilibrium distribution tends to return to equilibrium as a result of scattering processes. If τ is the same for all electrons over the Fermi surface, eqn (10.9) is reproduced.

With this same assumption, the classical expression for the thermal conductivity due to the electrons can be obtained from the Boltzmann equation in a derivation which is closely analogous to that for the phonon case. We therefore obtain

$$\frac{\kappa}{\sigma} = \frac{1}{3} \frac{c_e v^2 m_e}{e^2}.$$

If we now substitute the correct expressions for c_e and v from eqns (10.8) and (10.6) we obtain

$$\frac{\kappa}{\sigma T} = \frac{\pi}{3} \left(\frac{k_B}{e}\right)^2.$$

This ratio, which is very little different from the classical value derived by using completely wrong heat capacities and velocities, is called the ideal Lorenz number L_0, and has the numerical value $2 \cdot 45 \times 10^{-8}$ W-Ω deg^{-2}.

10.4. Conditions for the validity of the Wiedemann–Franz–Lorenz law

The derivation of the law assumes that the relaxation times, or mean free paths, appropriate to thermal and electrical conduction are identical. However, the form of the departure of the electron distribution from equilibrium due to an electric field is different from the departure produced by a temperature gradient. Fig. 10.4 showed how the Fermi surface is displaced by an electric field, but the boundary of the surface itself is only sharp at 0 K when every possible k value within its boundaries is occupied by an electron. At a finite temperature there are levels below E_F which are not full and levels above E_F which have some probability of being occupied. This lack of sharpness of the surface can be put into the illustration of the effect of fields by reducing the two-dimensional representation of the three-dimensional Fermi surface still further to one dimension, and representing by the vertical axis the probability of any energy level, or k value, being occupied. At 0 K the edges of the picture would be vertical and the curvature shown in Fig. 10.5a represents the situation at some finite temperature. As shown in § 10.3, the whole figure moves to the right in an electric field, but keeps its shape (if τ is the same for all electrons near the Fermi surface).

Thermal conductivity is normally measured under conditions which prevent an electric current from flowing, so that there is no net flow of electrons. The same number of electrons must therefore be flowing in unit

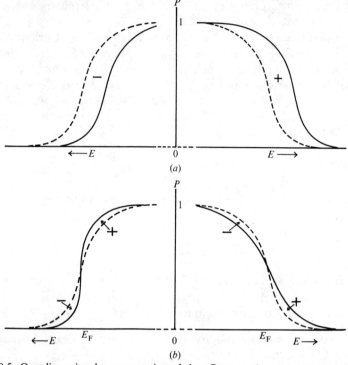

FIG. 10.5. One-dimensional representation of the effects on the probability P of finding electron energy levels near E_F occupied, in the presence of (a) an electric field and (b) a temperature gradient. The broken curves represent the undisplaced probabilities; the + and − signs indicate energy levels which become over-populated and under-populated respectively. To the right of the origin the values of k would be positive and they would be negative to the left.

time in opposite directions, and the heat current results from electrons at lower density travelling in the direction of the heat flow with greater energy than the higher-density stream flowing up the temperature gradient. In order to see how the distribution in the one-dimensional representation is changed by the gradient, we can use the solution of the Boltzmann equation for a fixed relaxation time (eqn (4.7)), which in terms of the electron occupation numbers gives for one dimension

$$\mathscr{F} - \mathscr{F}^0 = v\tau \frac{dT}{dz} \frac{\partial \mathscr{F}^0}{\partial T}.$$

If for simplicity we assume that v and τ are the same for all the effective electrons, which must be close to the Fermi level, and that the variation of E_F with temperature can be neglected, we find

$$\mathscr{F} - \mathscr{F}^0 = \pm \mathscr{B} \frac{\partial \mathscr{F}^0}{\partial T} \qquad (10.10)$$

where $\mathcal{B} = v\tau(\mathrm{d}T/\mathrm{d}z)$. If in Fig. 10.5b, energy is flowing to the right, the temperature gradient is negative and v is positive for electrons to the right of the centre and negative for electrons to the left. The positive sign in eqn (10.10) thus applies to electrons to the right. Now

$$\frac{\partial \mathscr{F}^0}{\partial T} = \frac{(1/T)\{(E-E'_\mathrm{F})/k_\mathrm{B}T\}\exp\{(E-E'_\mathrm{F})/k_\mathrm{B}T\}}{\{\exp(E-E'_\mathrm{F})/k_\mathrm{B}T+1\}^2}.$$

If we denote $(E-E'_\mathrm{F})/k_\mathrm{B}T$ by x'

$$\mathscr{F}-\mathscr{F}^0 = \pm\mathcal{B}\frac{x'\mathrm{e}^{x'}}{T(\mathrm{e}^{x'}+1)^2}.$$

At the Fermi level ($x' = 0$) the value of $\mathscr{F}$ is unchanged for electrons travelling in either direction, so that the patterns with and without a temperature gradient cross at these points. Far away from E_F ($|x'|$ large) there is again no change in $\mathscr{F}$; the exponentials in the numerator and denominator ensure this for large negative and positive x' respectively. Near E_F, for electrons travelling to the right, $\mathscr{F}$ is increased if x' is positive and decreased if x' is negative. For electrons travelling to the left the converse holds, as is illustrated in Fig. 10.5b.

We see that the out-of-equilibrium distributions are quite different for an electric field and for a temperature gradient, and this can lead to the relaxation rates appropriate to the two cases being different too. In order to produce electrical resistance by tending to restore the equilibrium distribution, a scattering process must take electrons from the right of Fig. 10.5a and carry them round to the left. If this is to be done in one process, it must be capable of making a drastic change in the electron wave-vector. On the other hand, to restore the thermally produced distribution back to equilibrium, either the same large wave-vector changes can occur, so that the electrons from the over-full levels fill up corresponding empty levels on opposite sides of the figure, or, by a very small change in wave-vector and energy, electrons from the over-full levels on the right can fill over-empty levels also on the right, and similarly on the left. As the range of energies over which the Fermi distribution changes from 1 to 0 is a few times $k_\mathrm{B}T$, the latter process only involves energy changes of the order of $k_\mathrm{B}T$ and correspondingly small wave-vector changes. We shall see that if scattering by the lattice (by phonons) is the dominant cause of resistance, then the WFL law breaks down when the dominant phonons have such small wave-vectors that they cannot bring about, in one collision, the large wave-vector changes needed to produce electrical resistance, but can yet bring about the small wave-vector changes effective in causing thermal resistance.

11
ELECTRON SCATTERING

To a first approximation the electronic thermal and electrical conductivities can be considered to be determined by a number of scattering processes, each of which gives rise to an identifiable resistivity. According to Matthiessen's rule (1862) these resistivities are additive. One component resistance which is always present is the 'ideal' resistance due to scattering of electrons by the lattice.

The scattering of electrons by phonons forms an essential part of any discussion of electrical conductivity and will not be treated in detail here. However, when considering the corresponding thermal conductivity it must be remembered that the departures from thermal equilibrium brought about by electrical and thermal fields are different.

In a metal, phonons as well as electrons can transport heat, and the phonon–electron interactions which limit the electronic conduction also limit the phonon conduction. Over wide temperature ranges, scattering of phonons by electrons is the major factor in determining the lattice conductivity of a metal.

11.1. Phonon–electron scattering

The calculation of the transition probability for an electron with wave-vector $\mathbf{k}_1$ to be scattered into state $\mathbf{k}_2$ as a result of an interaction involving a phonon of wave-vector $\mathbf{q}$ is discussed in, for example, Ziman (1960). In order to simplify the calculation of conductivity many approximations are usually made (such as the adiabatic approximation that the electron wave-functions keep pace with the ionic motions and the assumption of a spherical Fermi surface), but there are a number of difficulties which can be more or less ignored if we are content to restrict our interest to the relations between the various conductivities—the electrical conductivity, the electronic thermal conductivity, and the lattice conductivity—and their temperature dependences, and do not aspire to predict their absolute magnitudes.

The conditions which have to be fulfilled among electrons and phonons for an interaction to take place are similar to those which are fulfilled for phonon–phonon interactions:

$$E(\mathbf{k}_1) \pm E(\mathbf{q}) = E(\mathbf{k}_2) \tag{11.1a}$$

and

$$\mathbf{k}_1 \pm \mathbf{q} = \mathbf{k}_2 + \mathbf{g}. \tag{11.1b}$$

The positive signs correspond to absorption of a phonon by electron $\mathbf{k}_1$, and the negative signs to emission of a phonon. As for phonons, these conditions correspond to conservation of energy and 'momentum' (apart from the addition of a reciprocal lattice vector).

Eqn (11.1a) (with the positive sign) can be written as

$$\hbar\omega_q - \{E(\mathbf{k}_2) - E(\mathbf{k}_1)\} = 0. \tag{11.1c}$$

The maximum phonon energy $\hbar\omega_D$ is usually much smaller than the energy E_F of the effective electrons, so that $E(\mathbf{k}_2) - E(\mathbf{k}_1)$ is relatively small, and by using eqn (11.1b), with $\mathbf{g} = 0$, can be written as $(\partial E/\partial \mathbf{k})_{E_F} \cdot \mathbf{q}$ or $\hbar \mathbf{v}_F \cdot \mathbf{q}$. Eqn (11.1c) then becomes

$$\omega_q - \mathbf{v}_F \cdot \mathbf{q} = 0$$

or

$$\omega_q/q - (\mathbf{v}_F \cdot \mathbf{q})/q = 0.$$

In an electron–phonon N-process the component of the initial electron velocity in the direction of the phonon must thus be equal to the phase velocity of the lattice wave.

For absorption of a phonon the matrix element for the transition can be written as $\mathcal{N}_q^{\frac{1}{2}} G_q$, where G_q contains the strength of the phonon–electron interaction and is inversely proportional to $(\omega_q)^{\frac{1}{2}}$; G_q depends on the polarization vector of the phonon and on the scattering vector $\mathbf{k}_2 - \mathbf{k}_1$, the exact form of the dependence being determined by the model used. For a spherical Fermi surface and N-processes ($\mathbf{g} = 0$ in eqn (11.1b)) only longitudinal phonons interact with electrons, but for other shapes of Fermi surface and for U-processes transverse phonons also interact.

In order to see how the main qualitative features of the resistivities arise, we shall take the simplest model in which G_q is proportional to $\mathbf{k}_2 - \mathbf{k}_1$, and thus for N-processes is proportional to $\mathbf{q}$. U-processes are ignored and no distinction is made between different phonon polarizations. The rate of change of population of electrons $\mathbf{k}_1$ due to absorption of phonons $\mathbf{q}$ is proportional to $\mathcal{N}_q(q^2/\omega_q)\mathcal{F}_{k_1}(1-\mathcal{F}_{k_2})$, where the last two factors result from the Pauli principle, since the state $\mathbf{k}_1$ must be initially occupied and the state $\mathbf{k}_2$ initially unoccupied. There is a similar expression, with $\mathcal{N}_q + 1$ replacing $\mathcal{N}_q$ and with $\mathcal{F}_{k_1}$ and $\mathcal{F}_{k_2}$ interchanged arising from the reverse process in which a phonon $\mathbf{q}$ is emitted when electron $\mathbf{k}_2$ changes its state to $\mathbf{k}_1$.

When calculating the rate of change in the electron distribution it is usual to assume that the phonon distribution corresponds to thermal equilibrium (and vice versa). With this and other simplifying assumptions, including the absence of phonon dispersion (so that $\omega \propto q$), the rate of change in $\mathcal{F}_{k_1}$ is proportional to $\mathcal{N}_q^0 \omega_q$ and this must be summed over all possible values of q. In order to turn this sum into an integral the number of phonon modes available for the scattering which lie between q and $q+dq$ has to be inserted. For $\mathbf{k}_1$ fixed, $\mathbf{k}_2$ lies on a constant energy surface corresponding to an energy larger by $\hbar\omega$ than $E(\mathbf{k}_1)$ (if the phonon is absorbed). For a spherical Fermi surface, $\mathbf{k}_2$ lies on the surface of a sphere centred on the origin in k-space. For a given value of q, $\mathbf{k}_2$ also lies on the surface of a sphere of radius q with centre at $\mathbf{k}_1$. Interactions are thus possible which lead to values of $\mathbf{k}_2$ lying on the circle of intersection of these two spheres as shown in Fig. 11.1 (in two dimensions the intersection occurs at two points). The circumference of the circle of intersection is proportional to q, since $|\mathbf{k}_2|-|\mathbf{k}_1| \ll |\mathbf{q}|$, and the area between circles corresponding to q and to $q+dq$ is proportional to $q\,dq$, which is proportional to $\omega\,d\omega$ when there is no dispersion.

Before concluding that the relaxation rate is just proportional to $\int \mathcal{N}^0(\omega)\omega^2\,d\omega$, it is necessary to consider how effective the changes in electron populations are in restoring to thermal equilibrium the different distributions produced by an electric field and by a temperature gradient, as discussed in § 10.4.

The maximum change in electron energy which can be produced by interaction with a phonon is equal to the maximum available phonon energy. On the Debye theory this is $k_B\theta$ and is small compared with E_F (for example, in copper $k_B\theta/E_F \sim 4\times 10^{-3}$). The change in electron wave-vector can, however, be relatively large since the maximum phonon wave-vector q_D is comparable with k_F. On the Debye model the Brillouin

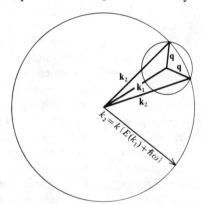

FIG. 11.1. Two-dimensional representation of possible electron–phonon interactions represented by

$$\mathbf{k}_1 + \mathbf{q} = \mathbf{k}_2$$
$$E(\mathbf{k}_1) + \hbar\omega = E(\mathbf{k}_2)$$

In three dimensions the circles become spheres which intersect over a circle. (Based on Klemens 1969.)

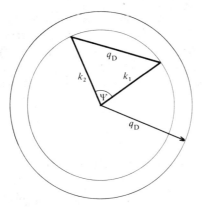

FIG. 11.2. The maximum change in direction, Ψ, of an electron on the Fermi surface produced by an N-process interaction with a phonon. It is assumed that the electrons are 'free' and the Brillouin zone is represented by the Debye sphere of radius q_D. Since the electron energy is hardly changed, $|\mathbf{k}_1| = |\mathbf{k}_2| = k_F$ and

$$\Psi_{max} = 2\sin^{-1}\left(\frac{q_D/2}{k_F}\right) = 2\sin^{-1}\left(\frac{2^{\frac{1}{3}}}{2}\right)$$

for one electron per atom.

zone for a crystal of N atoms would be a sphere containing N possible values of q. If each atom contributes one 'free' electron, the Fermi surface would have to accommodate $N/2$ values of k. The volume of the Brillouin zone would then be double the volume inside the Fermi surface, and for a spherical Fermi surface the ratio q_D/k_F would be $2^{\frac{1}{3}}$, as shown in Fig. 11.2. The maximum change in $\mathbf{k}$ on this simple model would be $2^{\frac{1}{3}}k_F$, and, if the small energy change is neglected, the maximum change in direction, Ψ, of k produced by a phonon would be $2\sin^{-1}(2^{\frac{1}{3}}/2)$, i.e. 79°. In a real solid both the Brillouin zone and the Fermi surface have quite different shapes, but we can assume that high-energy phonons can still produce drastic changes in $\mathbf{k}$ but little change in the electron energy.

When considering the effect of phonon–electron interactions on electrical conductivity it is necessary to know the extent to which a collision is capable of randomizing the electron momentum. The effectiveness of a collision from this point of view is proportional to $(1-\cos\Psi)$; it is a minimum when $\Psi = 0$ and a maximum when the direction of $\mathbf{k}_2$ is completely opposite to that of $\mathbf{k}_1$.

At high temperatures the dominant phonons have high values of q, so that electrons can be scattered over large angles and the mean value of $\cos\Psi$ can be taken as zero. Collisions are then equally effective in restoring to equilibrium the out-of-equilibrium distributions relevant to electrical and thermal conductivities.

At low temperatures the dominant phonons have energies $\sim k_B T$ and correspondingly small wave-vectors. In an N-process the electron wave-vector can only be changed comparatively little, and the angle between $\mathbf{k}_2$ and $\mathbf{k}_1$ is correspondingly small. For small q, the largest value of Ψ is q/k_F. Making the usual substitution $x = \hbar\omega/k_B T$ and neglecting dispersion, we have

$$\Psi = \frac{xk_B T}{\hbar v k_F} = \frac{xT}{\theta} \frac{k_B \theta}{\hbar v k_F} = \frac{xT}{\theta} \frac{q_D}{k_F}.$$

For a monovalent metal with a roughly spherical Fermi surface $q_D \sim k_F$, so that the maximum angular change is $\Psi \sim xT/\theta$. For small angles, $1 - \cos\Psi \sim \Psi^2/2$. When summing the relaxation rates for different q's to obtain the effective relaxation time for electrical conductivity, each term must therefore be weighted by the factor $\sim \frac{1}{2}x^2(T/\theta)^2$. However, this factor does not arise in the determination of the relaxation time effective for thermal conductivity, since even changes in electron energy of the order of $k_B T$ with no attendant change in direction suffice to take electrons from just below the Fermi level to just above it and vice versa. These seemingly minute adjustments to the distribution are all that are required to undo the departures from equilibrium produced by a temperature gradient.

The effective relaxation rates for thermal conductivity and electrical conductivity at low temperatures, where the angular effect is important, can be represented as follows:

Thermal conductivity　　　*Electrical conductivity*

$$\frac{1}{\tau_\kappa} \propto \int \mathcal{N}^0(\omega)\omega^2 \, d\omega \qquad \frac{1}{\tau_\sigma} \propto T^2 \int \mathcal{N}^0(\omega)\omega^2 x^2 \, d\omega$$

$$\propto T^3 \int_0^{x_{max}} \frac{x^2 \, dx}{e^x - 1} \qquad \propto T^5 \int_0^{x_{max}} \frac{x^4 \, dx}{e^x - 1}$$

At low temperatures the upper limit of integration is large, being θ/T, and the integrals tend to constant values. From this we deduce that $\tau_\kappa \propto T^{-3}$ and $\tau_\sigma \propto T^{-5}$, so that the conductivities should behave as $\kappa \propto T^{-2}$ and $\sigma \propto T^{-5}$ at low temperatures. Quite apart from the numerical coefficients, it is evident that the Wiedemann–Franz–Lorenz law must break down since $\kappa/\sigma T \propto T^2$ and this ratio tends to zero at 0 K. At high temperatures the integrals are proportional to x_{max}^2 and x_{max}^4 respectively, so that $\tau_\kappa \propto T^{-1}$ and $\tau_\sigma \propto T^{-1}$. In fact, the relaxation times also have the same magnitude so that the WFL law is obeyed, as we would expect for large-angle scattering. Laubitz and Cook (1972) have discussed the contribution to the high-temperature thermal resistivity of transitions across the 'thickness' of the Fermi surface.

The same interactions between electrons and phonons which change the populations of electron states also change the phonon populations and give rise to a phonon thermal resistivity additional to any of the other sources of resistance which occur in non-metallic crystals. In order to obtain the relaxation rate for phonons q, we assume that the electron system is in thermal equilibrium ($\mathscr{F}_k = \mathscr{F}_k^0$) and sum the net rate of change in $\mathcal{N}_q$ over all combinations of $\mathbf{k}_1$ and $\mathbf{k}_2$ which can interact with phonons $\mathbf{q}$. This leads to a relaxation rate proportional to ω_q, so that the corresponding contribution to the phonon resistivity $W_e^p \propto T^{-2}$ at low temperatures and is constant at high temperatures (these results follow from substituting $\tau(x) \propto (xT)^{-1}$ in the standard integral for thermal conductivity (eqn (4.9b))).

We thus have three quantities which are intimately related to one another, and in the ensuing discussion the following nomenclature will be adopted: the superscript e or p denotes the carrier concerned in the conduction, while the subscript e or p denotes the carrier which is producing the resistance to the current. The electrical conductivity and resistivity when limited by phonon scattering are thus written as σ_p^e and ρ_p^e (although here the superscript can be omitted since phonons do not carry electric charge), the electronic thermal conductivity and resistivity as limited by phonons are κ_p^e and W_p^e, and the phonon thermal conductivity and resistivity when limited by electron scattering are κ_e^p and W_e^p.

In order to consider the inter-relationships between the various resistivities the more accurate forms for them derived by a proper treatment of the same simplified model of a metal are given below. The three relations were first given by Bloch (1930), Wilson (1937), and Makinson (1938), respectively. More detail can be found in Wilson (1953) and Ziman (1960). The relations are as follows:

$$\rho_p^e = A\left(\frac{T}{\theta}\right)^5 J_5\left(\frac{\theta}{T}\right) \qquad (11.2)$$

$$W_p^e = \frac{A}{L_0 T}\left(\frac{T}{\theta}\right)^5 J_5\left(\frac{\theta}{T}\right)\left\{1 + \frac{3}{\pi^2}\left(\frac{k_F}{q_D}\right)^2\left(\frac{\theta}{T}\right)^2 - \frac{1}{2\pi^2}\frac{J_7(\theta/T)}{J_5(\theta/T)}\right\} \qquad (11.3)$$

$$W_e^p = \frac{A}{L_0 T}\left(\frac{\theta}{T}\right) J_5\left(\frac{\theta}{T}\right)\frac{\pi^2 n_a^2}{27\{J_4(\theta/T)\}^2} \qquad (11.4)$$

where

$$J_n\left(\frac{\theta}{T}\right) = \int_0^{\theta/T} \frac{x^n e^x}{(e^x - 1)^2}\,dx \quad \text{and} \quad A = \frac{3\pi \hbar q_D^6 (G')^2}{4e^2 m_e^2 n_c k_B \theta k_F^2 v_F^2}.$$

In these equations n_c is the number of unit cells per unit volume and n_a the number of electrons per atom. It must be emphasized again that electron–phonon U-processes are ignored in deriving these expressions, and the strength of the electron–phonon interaction is represented (apart from the $\mathbf{k}_2 - \mathbf{k}_1$ factor) by the constant G'.

The appearance of, for example, the integral denoted by J_5 in eqn (11.2) is not consistent with the expressions obtained previously for $1/\tau_\sigma$, since on integrating by parts

$$\int_0^{\theta/T} \frac{x^{n-1} \, dx}{e^x - 1} = \left[\frac{1}{n} \frac{x^n}{e^x - 1}\right]_0^{\theta/T} + \frac{1}{n} \int_0^{\theta/T} \frac{x^n e^x}{(e^x - 1)^2}.$$

For low temperatures (large θ/T) the term in square brackets is zero at both limits, so that the left-hand integral is just $J_n(\theta/T)/n$. For large values of θ/T the limiting values are $J_4(\infty) = 25.98$, $J_5(\infty) = 124.4$, and $J_7(\infty) = 5082$, and these limits are reached to within 1 per cent for θ/T equal to 11.5, 13, and 16, respectively.

For high temperatures we assume that x is always small so that e^x in the denominator is always $1 + x$. The left-hand integral is then $x_{\max}^{n-1}/(n-1)$, which is the same as $J_n(x)$ in the same limit.

The three terms which make up the electronic thermal resistivity can be considered as the effects of (a) large-angle scattering which produces a corresponding electrical resistivity which is related by the WFL law, (b) small-angle scattering which has no counterpart in the electrical resistivity, and (c) a 'correction' for the fact that, when the energy change is small, large-angle scattering can reverse the electron direction without helping to restore the equilibrium distribution. This arises when scattering occurs from one side of the Fermi surface to the other between regions for which the deviations from the equilibrium populations are similar (such as the right and left regions marked + in Fig. 10.5b).

11.2. The temperature variations of the resistivities

11.2.1. The 'ideal' electrical resistivity

At high temperatures $T \gg \theta$, $J_5(\theta/T) = \frac{1}{4}(\theta/T)^4$, so that from eqn (11.2)

$$\rho_P^e = A \left(\frac{T}{\theta}\right)^5 \frac{1}{4} \left(\frac{\theta}{T}\right)^4 = \frac{A}{4} \frac{T}{\theta}. \tag{11.2a}$$

At low temperatures, $T \ll \theta$, $J_5(\theta/T) = 124.4$, and

$$\rho_P^e = 124 \cdot 4 A \left(\frac{T}{\theta}\right)^5. \tag{11.2b}$$

11.2.2. The 'ideal' electronic thermal resistivity

At high temperatures $\theta/T \to 0$ and

$$\frac{J_7(\theta/T)}{J_5(\theta/T)} \to \frac{\frac{1}{6}(\theta/T)^6}{\frac{1}{4}(\theta/T)^4} \to \frac{2}{3}\left(\frac{\theta}{T}\right)^2 \to 0$$

ELECTRON SCATTERING

so that the second and third terms in the square brackets of eqn (11.3) are both zero. Then

$$W_p^e = \frac{A}{L_0 T}\left(\frac{T}{\theta}\right)^5 J_5\left(\frac{\theta}{T}\right) \quad (11.3a)$$

$$= \frac{\rho_p^e}{L_0 T}.$$

The WFL law is obeyed and W_p^e is constant.

At low temperatures J_7/J_5 reaches the constant value $5082/124 \cdot 4$, while the middle term in eqn (11.3) increases indefinitely and becomes the dominant contribution to W_p^e, which then becomes

$$W_p^e = \frac{A}{L_0 T}\left(\frac{T}{\theta}\right)^3 J_5\left(\frac{\theta}{T}\right)\frac{3}{\pi^2}\left(\frac{k_F}{q_D}\right)^2 \quad (11.3b)$$

$$= \frac{\rho_p^e}{L_0 T}\left(\frac{\theta}{T}\right)^2 \frac{3}{\pi^2}\left(\frac{k_F}{q_D}\right)^2$$

showing the large deviation from the WFL law at low temperatures for metals which are perfect enough for the 'ideal' resistivity to dominate down to sufficiently low temperatures.

On the model used W_p^e passes through a maximum for θ/T between 4 and 5 if k_F/q_D is taken as $\frac{1}{2}^{\frac{1}{3}}$, the value it would have for a spherical Fermi surface corresponding to one electron per atom in a Debye-type crystal lattice. This maximum would be 60 per cent greater than the limiting high-temperature resistivity. The existence of the maximum is not connected with the existence of the third term in eqn (11.3) and would in fact be greater without it.

The position of the predicted maximum in W_p^e is sensitive to the value of k_F/q_D and moves to higher temperatures (smaller θ/T) as this ratio decreases. For example, if $k_F/q_D = (\frac{1}{2} \times 0 \cdot 3)^{\frac{1}{3}}$ there is no maximum at all. On our simple model for a metal this corresponds to $0 \cdot 3$ electrons per atom.

11.2.3. The phonon thermal resistivity due to electron scattering

At high temperatures W_e^p from eqn (11.4) becomes

$$W_e^p = \frac{A}{L_0 T}\left(\frac{\theta}{T}\right) J_5\left(\frac{\theta}{T}\right)\frac{\pi^2 n_a^2}{(\frac{27}{9})(\theta/T)^6} = \frac{\rho_p^e}{L_0 T}\frac{\pi^2 n_a^2}{3}$$

so that, if only scattering by electrons were important in determining the phonon conductivity, this would be comparable with the electronic thermal conductivity. However, at the temperatures where this simple relation would hold, phonon–phonon U-processes are dominant in determining the phonon conductivity and this contribution to the high-temperature thermal conductivity of a metal is much depressed.

At low temperatures eqn (11.4) can be written

$$W_e^p = \frac{A}{L_0 T}\left(\frac{\theta}{T}\right) J_5\left(\frac{\theta}{T}\right) \frac{\pi^2 n_a^2}{27 \times 25 \cdot 9^2} = \frac{\rho_p^e}{L_0 T}\left(\frac{\theta}{T}\right)^6 \frac{\pi^2 n_a^2}{1 \cdot 8 \times 10^4}.$$

We can then compare the low-temperature phonon and electronic thermal conductivities:

$$\frac{W_e^p}{W_p^e} = \left(\frac{\theta}{T}\right)^4 \frac{\pi^4 n_a^2}{5 \cdot 4 \times 10^4 (k_F/q_D)^2}.$$

Since the simple representations of the low-temperature conductivities are only valid for $\theta/T \geqslant 10$, it can be seen that, for our simple model with $n_a = 1$ and $(k_F/q_D)^2 = 0 \cdot 63$, W_e^p is much greater than W_p^e, so that in a pure metal in which the ideal electronic thermal resistivity dominates the electronic contribution the thermal conductivity is almost entirely due to electrons.

11.3. Some corrections to the simple theory

No real metal has properties very similar to those which have been assumed in the simple theories of conductivity, so that close agreement between predictions and experimental observations cannot be expected. Improvements to the theory can be made by modifying the model to fit more closely the properties of particular metals, but such modifications will not be discussed here. There are, however, several corrections of a more general nature applicable to all metals, and some of these will now be outlined.

11.3.1. Electron–phonon U-processes

The expressions given in eqns (11.2), (11.3), and (11.4) for the various resistivities are derived by considering only electron–phonon N-processes. U-processes are possible, just as for phonon–phonon interactions, when the phonons have large values of q, but in addition U-processes can be brought about by phonons with vanishingly small q if the Fermi surface is sufficiently close to the zone boundary. A U-process for a spherical Fermi surface and a cubic Brillouin zone is illustrated in Fig. 11.3. The electron represented by k_2 is identical with electron k_2', and in directions such that k_1 is close to the zone boundary such a U-process can result from interaction with a very-low-energy phonon.

U-processes increase the number of ways in which electrons can be scattered and increase both the electrical and the electronic thermal resistivities. Estimates vary as to the magnitude of this increase but it is of the order of 2 for simple metals and can be much more. There is a further correction in the same direction when the normal departures from the Debye theory are taken into account.

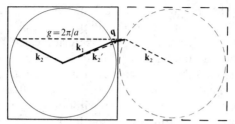

FIG. 11.3. Representation of an electron–phonon U-process $\mathbf{k}_1+\mathbf{q}=\mathbf{k}_2+\mathbf{g}$.

The appreciable maximum in the electronic thermal resistivity associated with the expression for W_p^e (eqn (11.3)) has not been observed (observation of small maxima will be discussed in § 12.1.2(a)). It arises from the presence of the second term in eqn (11.3), and occurs when this term is comparable with the first but is not yet dominant. The existence of U-processes requiring only very small phonon energies enhances the large-angle scattering down to lower temperatures and depresses the maximum.

If the Fermi surface is very close to the zone boundary or touches it, U-processes can occur at low temperatures. It was assumed, when only N-processes were considered, that at low temperatures, because of the factor $(1-\cos\Psi)$, scattering is very ineffective for electrical resistivity. Electrons can be reversed in direction in far fewer steps by U-processes than by an N-process, and this implies that the angular factor should not be so potent in reducing the electrical resistivity as the temperature decreases. The low-temperature electrical resistivity should therefore not be so low in comparison with the high-temperature electrical resistivity, or with the low-temperature electronic thermal resistivity in which the angular factor did not, in any case, appear.

An early calculation by Bardeen (1937) included U-processes and leads to an electronic thermal resistivity with practically no maximum (Ziman 1954). Only minor adjustments to this are needed to remove the maximum altogether.

11.3.2. Electron–electron scattering

The effect of electron–electron scattering has been ignored in deriving the expressions for electrical resistivity and electronic thermal resistivity. An interaction among electrons would have to involve four electron states. As for phonons, an N-process would not have any effect on the resistivity, but resistance is caused by U-processes in which the net electron wavenumber changes by a reciprocal lattice vector. There are, however, two reasons why the probability of such processes is small.

If a charge is placed in a sea of electrons, the electron concentration in the neighbourhood of the charge is distorted in such a way that the potential around the charge falls off much more rapidly with distance than it would in the absence of the surrounding mobile charges. As a result, the cross-section for electron–electron scattering calculated, say, on the Rutherford scattering model, is quite small ($\sim 10^{-19}\,\text{m}^2$ for ordinary metals). In a real metal the effective scattering cross-section is still smaller because of the Pauli exclusion principle. In order that conservation of energy and total wave-vector (with or without **g**) may be satisfied in a collision between two electrons, with the final two states accessible, the combinations of states are severely restricted. This reduces the cross-section (assuming a spherical Fermi surface) by a further factor of $\sim(k_B T/E_F)^2$, which at room temperature is $\sim 10^{-4}$ (a simple derivation of this factor is given by Kittel (1971)). The electron mean free path at room temperature is then $\sim 10^{-6}\,\text{m}$, which is at least 10 times larger than the mean free path limited by electron–phonon interactions (10^{-7}–$10^{-8}\,\text{m}$).

The scattering by electron–electron interactions decreases at low temperatures but it can become the dominant resistive process for exceptionally perfect metallic specimens. Even if it is not dominant at low temperatures, it can sometimes be separated out by analysis of the measured total thermal resistivity.

These conclusions do not necessarily hold for more complicated metals, and most of the relevant discussion has referred to transition metals. It is thought that most of the conductivity is due to s-electrons, but that appreciable resistance can be caused as a result of their scattering by d-electrons. The T^2 term in the scattering cross-section then appears as a contribution to ρ^e proportional to T^2 and to W^e proportional to T. Since these contributions die out with decreasing temperature more slowly than ρ_P^e and W_P^e, they might be observable in pure transition metals at low temperatures. However, the resistivities caused by phonons are enhanced by the presence of additional states (in the d-band) into which the conduction electrons can be scattered, and the resistances caused by electron–electron scattering represent a smaller fraction of the actual resistances than might have been thought at first.

11.3.3. Electron–phonon scattering when the electron mean free path is small

The results given for the electron–phonon scattering rate are only valid if the electron mean free path is comparable with or greater than the phonon wavelength. As discussed by Ziman (1960), this implies that an electron has to sample all phases of the lattice wave with which it is interacting in order that the assumptions of the adiabatic approximation should be valid. This condition can be expressed as $ql^e > 1$.

In order to see whether this condition imposes any constraint on the theory in real cases, we can consider numerical values for copper. The electron mean free path in copper with resistivity $\rho\,\Omega\,m$ is $6\cdot 5 \times 10^{-16}/\rho$ m. For fairly pure copper the residual resistivity ρ_0 may be $\sim 5 \times 10^{-11}\,\Omega$ m, so that $l^e \sim 10^{-5}$ m. In copper longitudinal lattice waves have wavelengths greater than this mean free path unless the frequency is more than 5×10^8 Hz. This means that the standard perturbation theory can break down for ultrasonic waves propagating in pure copper at low temperatures.

The lower frequency limit for the validity of the condition $ql^e > 1$ is shifted to higher frequencies for less pure copper or for higher temperatures where the electrical resistivity has risen above ρ_0.

In copper the phonons dominant in heat conductivity have wavelengths given by $\lambda_{max} \sim (300/T) \times 10^{-10}$ m, and even at helium temperatures this is much less than l^e for pure copper. However, if the copper contains a relatively small amount of impurity which increases ρ_0 to $\sim 10^{-8}\,\Omega$ m, then l^e becomes similar to λ_{max} at helium temperatures. For less pure metal l^e is less than λ_{max} at higher temperatures.

Pippard (1955) developed a theory of the attenuation of ultrasonic waves in metals, and this theory is applicable over the entire range of values of ql^e, giving the same result as perturbation theory under the conditions for which the latter holds. On Pippard's theory, the electron–phonon interaction is reduced for both longitudinal and transverse waves for small values of ql^e. The attenuation of longitudinal waves reaches the perturbation-theory behaviour for $ql^e \sim 5$, when it becomes proportional to q, but for transverse waves it becomes independent of q and inversely proportional to l^e for large values of ql^e.

The effect of the reduced electron–phonon interaction would hardly be detectable in electron transport. In pure metals the phonons of small enough wave-vector for $ql^e < 1$ are not contributing appreciably to electron scattering, while, if the electron mean free path is reduced to involve phonons with higher q in the inequality, scattering of electrons is predominantly by the imperfections which have been introduced to lower the electron mean free path. On the other hand, in even quite impure metals and alloys at only moderately low temperatures, the dominant effect in determining the lattice thermal conductivity is scattering by electrons. The change in the electron–phonon interaction then leads to a change in the lattice conductivity as the purity of the metals is altered, even though this conductivity may be entirely determined by scattering by electrons.

There have been several discussions of the application of Pippard's theory to lattice conductivity. As these are based on the free-electron model, we should not expect detailed agreement with experiment. Fig. 11.4 shows the ratio $l_e^p/(l_e^p)_{pure}^{long}$ of the phonon mean free path to the value for longitudinal phonons in the pure metal as a function of ql^e calculated by Lindenfeld and Pennebaker (1962). In this figure the ratio $l^{trans}/l_{pure}^{long}$ is

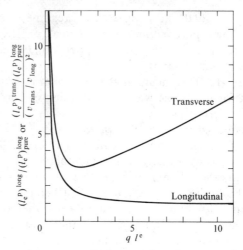

FIG. 11.4. The ratio of the phonon mean free path in an impure metal, l_e^p, to the mean free path for longitudinal phonons in the pure metal, $(l_e^p)_{\text{pure}}^{\text{long}}$, as a function of ql^e, the product of phonon wave-number and electron mean free path; $l^{\text{trans}}/l^{\text{long}}$ is divided by the square of the velocity ratio. (After Lindenfeld and Pennebaker 1962.)

divided by the square of the velocity ratio. The corresponding lattice conductivity was calculated on the assumptions of a Debye-type solid, with independent contributions from transverse and longitudinal modes. Zimmerman (1959) had previously deduced the behaviour of the conductivity on the assumption that Pippard's expression for the attenuation of longitudinal modes also applied to transverse modes.

The analysis of Lindenfeld and Pennebaker showed that, at temperatures low enough for l^e to be independent of temperature, the trend should be for the lattice conductivity to be linear in temperature and greater for less pure material for low Tl^e, and to increase more rapidly with temperature for large Tl^e and then to be greater for purer material. At higher temperatures, where l^e decreases, the behaviour of the conductivity is not so easily characterized; if, for example, Tl^e remained constant, κ_e^p would be proportional to T^2, but its magnitude would be different from the lattice conductivity of the pure metal.

11.4. Scattering by defects

In simple metals all the electrons effective in conduction have small wavelengths (a few tenths of 1 nm). They are contained within a narrow band near the Fermi level and there is little change in their energy with temperature. The scattering cross-section of static lattice defects for electrons, unlike phonons, is effectively the same for all electrons. This means that the electrical resistance due to defects is independent of temperature and that the electronic thermal resistivity is inversely proportional to the temperature—the scattering is generally elastic and provides a sufficiently large change in electron wave-vector to be equally effective for thermal and electrical conductivity. Calculations which have been made of the scattering cross-sections for various types of defect are equally valid for electronic thermal conductivity and electrical conductivity. The corresponding contribution to the electronic thermal resistivity can be deduced from the electrical resistivity by using the WFL law, and the calculations will not be discussed here.

In discussing conduction by phonons it was necessary to consider in some detail how different scattering mechanisms combine to produce the observed conductivity. This difficulty arises because most scattering processes are frequency dependent and a wide range of phonon frequencies is concerned in heat conduction. In addition, N-processes transfer energy among the modes so that phonons cannot be considered to carry heat independently. To a good approximation this problem does not arise in electrical conductivity, because of the small spread in electron wave-vectors, and for the same reason it is usually neglected in electronic thermal conductivity.

There can, however, be slight departures from simple additivity of resistances—Matthiessen's rule in the case of electrical conductivity—due to the energy-dependence of the scattering of electrons by phonons. In the case of electronic thermal resistivity at low temperatures we are concerned with small changes across the Fermi surface, and the variations of relaxation rate with energy for electron–phonon scattering can lead to deviations from Matthiessen's rule. The deviations in the case of electrical resistivity are much smaller, because the resistance is not very sensitive to small changes in electron wave-vector.

Sondheimer (1950) discussed deviations from Matthiessen's rule for thermal and electrical conduction by electrons on the Bloch model, and from the tables he gives one can deduce the orders of magnitude involved. If we determined the electrical resistance to be ascribed to impurities from the limiting low-temperature resistivity, then according to Matthiessen's rule this electrical resistance would have to be added to that of the pure metal at all temperatures. In the case of electronic thermal conductivity it would be necessary to determine the low-temperature limiting value of $W^e T$, and then this limiting value divided by the appropriate T would be the resistivity to be added to the ideal resistivity at any other temperature. According to Sondheimer, if the impurity resistance were sufficient to increase the electronic thermal resistance in the temperature region of $\theta/6$ to $\theta/5$ by 50 to 100 per cent of the ideal value, then in fact the total resistivity would be about 3 per cent more than would be expected by simple addition of the impurity resistance to the ideal resistance. At these temperatures the ideal electrical resistivity is relatively less than the ideal electronic thermal resistivity, because of the breakdown of the WFL law, so that the same amount of impurity contributes a relatively larger fraction of the total electrical resistivity. The departure from Matthiessen's rule for the electrical resistivity in this particular case is less than 0·6 per cent.

For both electronic thermal conductivity and electrical conductivity such deviations from Matthiessen's rule would be difficult to detect. Although the effect is larger for thermal conductivity, measurements cannot be made as accurately as measurements of electrical conductivity, and the deviations occur under conditions when the lattice conductivity is not completely negligible.

There are more complex processes which can give larger deviations from Matthiessen's rule, but these will not be discussed here.

12
THERMAL CONDUCTIVITY OF METALS AND ALLOYS

THE thermal conductivity of metals and alloys is of great practical importance, and for many purposes values can be deduced with sufficient accuracy from a knowledge of the electrical conductivity by using the Wiedemann–Franz–Lorenz law. For very pure metals the ideal resistivity, due to scattering of electrons by phonons, is dominant and the law breaks down at intermediate temperatures, as discussed in § 11.1. The thermal conductivity would be overestimated by direct application of the law. On the other hand, in alloys the impurity scattering of electrons can be so great that the electronic thermal conductivity is suppressed enough for the lattice component to represent a considerable contribution. Since this component has no counterpart in electrical conductivity, the thermal conductivity would be underestimated by relying on the electrical conductivity and the WFL law.

If appreciable accuracy is required in estimating the thermal conductivity of a metal or alloy from its electrical conductivity, compilations of values should be consulted in order to judge how great the departures from the law are for typical cases. The publications of the Thermophysical Properties Research Center (Touloukian 1970) contain graphs of thermal conductivity over a wide range of temperature, while Childs, Ericks, and Powell (1973) give results for temperatures from 300 K downwards, with an indication of which other properties have been measured on the specimens. Tables and graphs of the Lorenz ratio $\kappa/\sigma T$ for temperatures below 300 K are provided for technically important metals and alloys by Hust and Sparks (1973).

In this chapter the electronic and lattice thermal conductivities and their relationship to one another will be discussed. The behaviour of the Lorenz ratio $\kappa/\sigma T$ indicates how the relation between electrical conductivity and electronic thermal conductivity varies with temperature and with impurity scattering. Other aspects of the scattering of electrons are of great interest, but those which can be studied by measuring electrical conductivity will not be treated in any detail here. There is a brief section on superconductors.

12.1. Pure metals

Although the electronic component is dominant in the measured thermal conductivity of a fairly pure, simple metal, it is necessary to have some idea of the magnitude of the lattice component before concluding that this can be neglected.

In many alloys the lattice component may represent an appreciable fraction of the total conductivity, and the easiest way of estimating the lattice conductivity in a pure metal is by extrapolating the values deduced from experiments on alloys. Such experiments will be discussed in § 12.2.2, but an indication of the results to be expected will be given here before the electronic contribution is analysed.

12.1.1. Estimates of the lattice component

It was shown in § 11.2.3 that, on the simple theory, the lattice thermal conductivity would be about one-third of the electronic component at high temperatures if only phonon–electron scattering were important in determining both contributions. Although at high temperatures this scattering is dominant for electrons, the lattice conductivity is in general determined by phonon–phonon U-processes and the simple 3:1 relation does not hold. The resistivity due to U-processes increases with increasing temperature, while the electronic thermal conductivity is roughly constant at high temperatures.

At low temperatures the lattice resistivity due to scattering by electrons varies as T^{-2}, while the electronic thermal conductivity is linear in temperature. The lattice component is thus very small at sufficiently high and at sufficiently low temperatures, but may be significant at intermediate temperatures.

For an idea of the order of magnitude of the lattice conductivity κ^p, eqn (7.3a) can be used to determine the resistivity W_p^p when only U-processes are important. Although the conductivity will vary more rapidly than $1/T$ at sufficiently low temperatures, this change in behaviour has been neglected for the present, since it will only occur at temperatures such that scattering of phonons by electrons is becoming dominant. The lattice resistivity W_e^p, determined by electron scattering, can be estimated from eqns (11.3) and (11.4). Table 12.1 shows the result of such a calculation for copper. Since the magnitudes of both contributions to the thermal resistivity are extremely uncertain, they have been added together to give W^p, without any concern for the niceties of the theory of the lattice conductivity determined by several processes acting together.

It can be seen that the lattice component becomes a very small fraction of the measured thermal conductivity at both high and low temperatures, and only rises to about 3 per cent of the total between 50 and 200 K. A similar calculation for platinum, on the crude assumption that there is one free electron per atom, indicates that the lattice component might reach about 10 per cent of the total between 50 and 200 K. Work on alloys

METALS AND ALLOYS

TABLE 12.1

Estimation of the lattice conductivity in pure copper and its contribution to the total thermal conductivity

Temperature (K)	W_e^p from eqns (11.3) & (11.4) (m K W^{-1})	W_p^p from eqn (7.3a) (m K W^{-1})	W^p (m K W^{-1})	κ^p (W m^{-1} K^{-1})	κ_{meas}	$\kappa^p/\kappa_{\text{meas}}$
300	8×10^{-3}	$1\cdot8 \times 10^{-1}$	$1\cdot9 \times 10^{-1}$	5	400	0·01
200	7×10^{-3}	$1\cdot2 \times 10^{-1}$	$1\cdot3 \times 10^{-1}$	8	410	0·02
100	6×10^{-3}	6×10^{-2}	$6\cdot6 \times 10^{-2}$	15	500	0·03
40	9×10^{-3}	$2\cdot4 \times 10^{-2}$	$3\cdot3 \times 10^{-2}$	30	1600	0·02
10	4–8×10^{-2}†	6×10^{-3}	4–8×10^{-2}	12 to 25	18000‡	0·001 or less

† At this temperature it is difficult to determine W_p^e, the ideal electronic thermal resistivity, from experiment.

‡ At this temperature the thermal conductivity is very sensitive to the degree of perfection of the specimen.

which suggests that these orders of magnitude are correct will be discussed in § 12.2.2. Here it will suffice to remark that at high temperatures the ratio $\kappa/\sigma T$ for copper is close to the theoretical value L_0, suggesting that the thermal conductivity measured is almost entirely due to electrons, while at corresponding temperatures (scaled according to Debye temperature θ) $\kappa/\sigma T$ for platinum is appreciably greater than the value for copper, suggesting that part of the conductivity is contributed by carriers which do not contribute to the electrical conductivity.

12.1.2. The electronic component

12.1.2(a). The ideal resistivity determined by electron–phonon interactions. The feature of the temperature dependence of the thermal conductivity of a pure metal predicted by eqn (11.3) which is easiest to look for is the minimum at $\sim 0\cdot2\theta$. At higher temperatures the conductivity should be independent of temperature and at lower temperatures it should rise rapidly. On the Bloch model this conductivity minimum would be nearly 40 per cent below the limiting high-temperature conductivity if n_a is taken as 1. It has been realized for some time that no minimum of this extent is observed experimentally, and the usual explanation is that the effect of electron–phonon U-processes is to depress the high-temperature conductivity so much that the transition from the high-temperature behaviour to that at low temperatures, where U-processes have become much less important, is represented by a monotonic increase in conductivity. Fig. 12.1 shows schematically how U-processes could bring about this changed behaviour.

There is evidence, however, that a very shallow minimum, representing

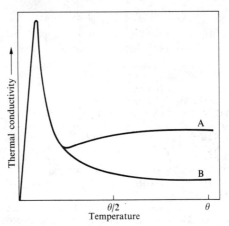

FIG. 12.1. Schematic representation of the action of U-processes in masking the conductivity minimum predicted by simple theory. For electron–phonon N-processes and one electron per atom, the minimum would occur at a temperature of $\sim 0\cdot 2\theta$ (curve A). The U-processes depress the high-temperature conductivity so much that there is a monotonic rise from high temperatures to the maximum (curve B). (From Klemens 1969.)

a dip of 3 to 5 per cent, occurs in the conductivity of sodium (Cook, van der Meer, and Laubitz 1972) and of aluminium (see Powell 1969). For sodium the minimum occurs near 70 K, which is just under half the Debye temperature θ_0 deduced from low-temperature specific-heat measurements (153 K), while for aluminium it occurs at about 180 K, which is again not much less than half θ_0 (426 K). Cook et al. have pointed out that the Debye temperature corresponding to longitudinal lattice vibrations alone would be ~ 260–300 K for sodium, so that the observed minimum actually occurs at $\sim \theta/4$ appropriate to those phonons which are assumed, on the Bloch model, to interact with electrons. However, another value of θ, that appropriate to scattering of electrons by phonons in the absence of U-processes, can be deduced by comparing the high-temperature and low-temperature ideal electrical resistivities (eqns 11.2a and 11.2b). Elimination of the constant A yields

$$\left(\frac{\rho_p}{T^5}\right)_{\text{low }T} = \frac{497\cdot 6}{\theta_R^4}\left(\frac{\rho_p}{T}\right)_{\text{high }T}$$

For both sodium and aluminium the values of θ_R appropriate to electrical resistivity are not very much greater than the values derived from specific-heat measurements.

From the absolute magnitudes of thermal and electrical conductivities at high temperatures it appears that U-processes account for a large part of the resistivities. The position of the minimum at $\sim 0\cdot 2\ \theta$ was derived from a theory which neglected U-processes, and the exact position is in any case sensitive to the effective number of electrons per atom. It therefore seems that the observed minima are not those predicted by eqn (11.3), but probably result from U-processes not being quite strong enough to lower the high-temperature conductivity sufficiently to suppress the minimum. In the absence of U-processes there would be a minimum of 2 per cent at just under $\theta/2$ if the effective number of electrons per atom were 0·35. Such a low value for sodium is not consistent with other properties of the metal.

For the monovalent metals the value of the ratio $\kappa/\sigma T$ increases with temperature and is close to the ideal value L_0 at room temperature or by about 100 °C, but it is difficult to discuss its behaviour at still higher temperatures because measurements of thermal conductivity become progressively more difficult. Cook *et al.* found that the thermal conductivity of sodium decreased above about 160 K, but this reflected the fact that the electrical resistivity was increasing more rapidly than the temperature, and $\kappa/\sigma T$ continued to rise until the melting point intervened. This departure from constancy in κ (or from proportionality of ρ to T) can be ascribed to variations with volume and temperature in the electron–phonon coupling constant (which affects the constant A) and in the effective value of θ. Just before melting a rapid increase in the concentration of lattice defects would be expected, and these would provide additional scattering of electrons.

At the lowest temperatures, where lattice defects dominate the scattering of electrons, the thermal conductivity becomes proportional to the temperature and the ratio $\kappa/\sigma T$ is indeed L_0. If we assume that electronic thermal resistivities are additive (the thermal equivalent of Matthiessen's rule), the ideal thermal resistivity at low temperatures can be found by subtracting $\rho_0/L_0 T$ from the measured resistivity to remove the effect of lattice defects. If the thermal and electrical conductivities are not determined on the same specimen using the same contacts, uncertainties in the exact specimen dimensions in the two experiments can render this method slightly inaccurate, especially at temperatures where the defect resistance is a predominant fraction of the measured resistance. It is then better to assume that the electronic thermal resistivity obeys the relation

$$W^e = W^e_d + W^e_p = a/T + bT^n.$$

Since n is expected to be 2 at low temperatures, a plot of $W^e T$ against T^3 should give something like a straight line. There are, in fact, a few metals for which this does result in a straight line, but for others there are deviations, and in a few extreme cases the value of n at low temperatures is as great as 3 (in these metals the variation in electrical resistivity is generally slower than T^5).

White and Woods (1959) list the powers of T which fit the ideal thermal resistivity at low temperatures for the transition metals which they measured, and also include values for sodium and the noble metals. For five out of the 22 metals it seems necessary to invoke a temperature dependence greater than 2·6, and there are two metals for which the power of T is marginally less than 2·0. It can be deduced from eqns (11.3a) and (11.3b) that in the region where the T^2 dependence should hold (below $\sim 0\cdot 1\theta$) the relation between the low-temperature ideal electronic thermal resistivity and the high-temperature limiting value would be

$$\left(\frac{W_p^e}{T^2}\right)_{\text{low}\,T} = 95 n_a^{\frac{2}{3}} \frac{W_\infty^e}{\theta^2} \qquad (12.1)$$

on the simple assumptions of the Bloch model, and of spherical Fermi surface and Brillouin zone. The numerical constant is smaller according to a higher-order variational calculation by Sondheimer (1950) and to a direct solution of the Boltzmann equation by Klemens (1954). The latter obtained a value of 64.

MacDonald, White, and Woods (1956) found that for the alkali metals the ratio

$$\frac{(W_p^e/T^2)_{\text{low}\,T}}{(W_\infty^e/\theta^2)_{\text{high}\,T}}$$

ranged from 13 (for sodium) to 23 (for caesium). For the noble metals White and Woods (1959) give the values of ~ 10 (silver and copper) and 14 (gold), and the values for the transition metals lie between 9 and 24. The exact values for the ratio depend on how the value of θ is chosen. The specific-heat values or those derived from the temperature dependence of electrical resistivity could be used, but even then the values appropriate to temperatures near 0 K or to a range of temperature would only differ slightly (see Meaden 1966). These various possibilities only change the range of values slightly without bringing any discernible pattern into the results. Also, if W_p^e is not proportional to T^2 the resistivity ratio depends on the temperature.

If the difference between the theoretical values for the ratio and the experimental values is ascribed to U-processes depressing the high-temperature conductivity (making $W_{p\infty}^e$ larger than is given by the simple theory), then this increase must in general be by a factor of about 4. We would not expect this factor to be at all the same for very different metals, nor could it be expected that the effective number of conduction electrons per atom would be similar. It is thus strange that there is so little variation in the value for the resistance ratio.

Another way of seeing to what extent the basic theory can account for observations derives from combining eqns (11.2b) and (11.3b) for the electrical and electronic thermal resistivities at low temperatures. If we again apply Klemens's correction of 0·68 to the standard expressions, take L_0 as the theoretical value, and make the usual simplifying assumptions already mentioned, we obtain

$$(W_p^e T^3)_{\text{low } T} = 5 \cdot 3 \times 10^6 n_a^{\frac{2}{3}} (\rho_p^e \theta^2)_{\text{low } T}.$$

MacDonald *et al.* (1956) tabulate values for the alkali metals of $(W_p^e T^3)_{\text{low } T}/(\rho_p^e \theta^2)_{\text{low } T}$, which range between $1 \cdot 7 \times 10^6$ (for sodium) down to $2 \cdot 9 \times 10^5$ (for lithium). Again W_p^e at low temperatures, which is little affected by U-processes, is being compared with a quantity (the electrical resistivity at low temperatures) which can be appreciably determined by U-processes (see § 11.3.1). It is thus to be expected that this resistivity ratio will be smaller than is given by simple theory.

Table 12.2 gives values for the two resistivity ratios which have been discussed (the first divided by 64 and the second by $5 \cdot 3 \times 10^6$. On the simple theory both would be $n_a^{\frac{2}{3}}$ (using Klemens's numerical correction). If U-processes increase the high-temperature thermal resistivity and the low-temperature electrical resistivity by the same amount, both ratios would still be equal, but less than $n_a^{\frac{2}{3}}$. For many metals the ratios are nearly the same, but what is more surprising is the relatively small variation among metals for which the effective number of conduction electrons is so different.

TABLE 12.2

Ratios of low-temperature electronic thermal resistivity to high-temperature electronic thermal resistivity and to low-temperature electrical resistivity†

Metal	Debye θ (K)	Temperature (K)	$\dfrac{(W_p^e/T^2)_{\text{low } T}}{64(W_p^e \infty/\theta^2)_{\text{high } T}}$	$\dfrac{(W_p^e T^3)_{\text{low } T}}{5 \cdot 3 \times 10^6 (\rho_p^e \theta^2)_{\text{low } T}}$
Li	360	‡	0·23	0·05
Na	160	‡	0·20	0·32
K	98	‡	0·28	0·12
Rb	61	‡	0·34	0·06
Cs	44	‡	0·36	0·06
Cu	320	20	0·17	0·17
Ag	220	20	0·16	0·15
Au	185	10	0·22	0·11
Pt	225	20	0·28	0·17
Fe	400	30	0·20	0·13
Rh	350	30	0·1	0·13
Zr	250	20	0·22	0·11

† Both ratios are expressed in such a way that they should be $n_a^{\frac{2}{3}}$ on the simple theory.
‡ At low temperatures for these metals $W_p^e \propto T^2$ and $\rho_p^e \propto T^5$, so that the ratios shown are independent of temperature.

Whatever the reasons for the variations in the two ratios, it does look as though a reasonable estimate of the ideal low-temperature electronic thermal resistivity can be derived from the limiting high-temperature value by using the simple theory with a numerical constant of about 15 in eqn (12.1).

12.1.2(b). Electron–electron scattering. As was mentioned in § 11.3.2 the mean free path for electron–electron scattering is very large, so that such processes can only have a noticeable effect on the conductivity in situations when the mean free path for all other scattering processes is also very large. Measurements on extremely pure metals are therefore required to show up any effect.

The scattering among electrons which lie within one band only leads to resistance if the scattering is by electron–electron U-processes. In simple metals the ideal electrical resistivity derived from experiment does not depart enough from proportionality to T^5 at low temperatures for it to be necessary to invoke such scattering. In transition metals, however, N-processes between s- and d-electrons can take an electron from a mobile s-state to a d-state of low mobility and thus reduce the current flow. In a number of transition metals the low-temperature variation of the electrical resistivity does decrease, and Garland and Bowers (1968) found an appreciable T^2 term in the electrical resistivity of aluminium and indium at helium temperatures. In indium the T^2 component exceeded the T^5 component at temperatures below 3 K. In the same set of experiments they found no such term for sodium and potassium. For indium the T^2 term was the same in specimens of different purities, while the T^5 term increased with increasing impurity, which was taken to indicate a departure from Matthiessen's rule for the addition of the electrical resistances due to impurities and to scattering by phonons.

It is more difficult to disentangle an electronic thermal resistance due to electron–electron scattering since its contribution is proportional to T, which is not very different from the T^2 variation of the resistivity due to scattering by phonons. At low enough temperatures, however, it might be distinguishable from the impurity resistivity which is proportional to T^{-1}. White and Tainsh (1967) analysed their measurements on nickel below 20 K and represented the resistivities by

$$\rho = a + bT^2$$

$$WT = c + dT^2$$

where a and c are constants; b and d are effectively constants but they seemed to change below 4–5 K, while staying in the same ratio of $1 \cdot 0 \times 10^{-8}$ W Ω K^{-2}. The constants a and c are in the ratio of the ideal Lorenz number L_0, as is expected for resistivities due to imperfections.

White and Tainsh also find that $\rho_i/\bar{W_i}T$ remains constant at 1×10^{-8} up to ~ 50 K, although at such a temperature neither ρ_i nor W_iT is proportional to T^2 because lattice scattering is becoming dominant.

Herring (1967) showed that, with certain assumptions, the Lorenz number relating electrical and electronic thermal resistivities due to electron–electron scattering should be lower than the standard value L_0 and gives the value $1\cdot 58\times 10^{-8}$. The value suggested by White and Tainsh is two-thirds of this.

Waldorf, Boughton, Yaqub, and Zych (1972) measured the thermal and electrical conductivities of very pure gallium, but found no evidence of a T^2 contribution to ρ or to WT, both of which were well represented by constants and terms in T^3, identified as the contributions from imperfection and phonon scattering. The Lorenz number corresponding to the measured conductivities approaches L_0 as $T \to 0$ K, but the Lorenz number corresponding to the ideal, temperature-dependent resistivities only decreases slowly below 4 K, where its value is $0\cdot 4\times 10^{-8}\,\text{W}\,\Omega\,\text{K}^{-2}$. From the simple theory we would expect the Lorenz number corresponding to the ideal resistivities to be given by $L^e \sim 8n_a^{-\frac{2}{3}}L_0T^2/\theta^2$, and for gallium at $3\cdot 2$ K ($\theta/100$) this is $\sim 2\times 10^{-11}\,\text{W}\,\Omega\,\text{K}^{-2}$, which is nothing like the experimental value even taking into account the discrepancy of a factor of about 4 in the relation between low-temperature thermal and electrical conductivities.

Waldorf *et al.* proposed quite a different explanation for the similar temperature dependences of ρ and WT at low temperatures for gallium. They suggested that the Fermi surface may have such great curvature that the small-angle scattering which is dominant at low temperatures can in fact produce large changes in the direction of the electron velocity. The angular factor introduced into the usual expression for electrical resistivity, which reflects the reduced effectiveness of scattering by low-energy phonons, should not therefore appear. The Lorenz number should then stay at about the ideal value, since this angular factor is not, in any case, involved in thermal resistivity.

12.2. Alloys

Most alloys used commercially contain a fair amount of impurity to impart the desired mechanical and thermal qualities. This depresses the electron thermal conductivity, but has generally little effect on the lattice component which is mainly determined by phonon–phonon U-processes and by electron–phonon interactions. The latter is, certainly, affected by the electron mean free path and thus by the defect scattering of electrons, but this dependence is weaker than the influence of defects on the electronic component of the thermal conductivity itself.

The lattice contribution to the total thermal conductivity in alloys can be determined more accurately than in metals, not only because it forms a larger proportion but because the electronic thermal conductivity can be estimated with more certainty from the electrical conductivity. The simplest way of determining the lattice conductivity in a metal is thus by measurements on alloys with different compositions and extrapolation of the deduced lattice conductivity to zero concentration of impurity.

12.2.1. *The electronic contribution*

In pure metals the experimental values of the electronic Lorenz number $\kappa^e/\sigma T$ should approach the ideal value L_0 at sufficiently high temperatures, but ordinary temperatures are often not high enough for it to be assumed that the ratio really is L_0. At low temperatures $\kappa^e/\sigma T$ falls below L_0 and for a very pure metal may reach a small fraction of L_0. In the limit of low temperatures as $T \to 0$ K, L^e rises again to reach L_0, but the temperature at which it does so differs from metal to metal and also depends on purity.

In an alloy the resistance due to imperfections is increased by the random nature of the atomic arrangement, and the temperature range over which this resistance dominates L^e is moved to higher temperatures. If this range is pushed up to over $\sim 0.5\theta$, L has not had a chance to fall much below L_0 from the high-temperature side before it rises again at low temperatures because the imperfection resistance has become dominant. Fig. 12.2 shows how L^e varies with T/θ for perfect and imperfect metals,

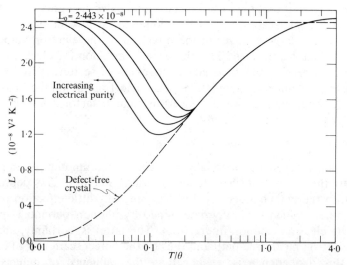

FIG. 12.2. The Lorenz ratio $\kappa^e/\sigma T$ for an ideally perfect metal and for specimens with imperfections plotted as a function of the ratio of temperature to the Debye characteristic temperature. (After Hust and Sparks 1973.)

according to Hust and Sparks (1973), and it can be seen that it has not fallen far below L_0 by $\theta/2$. If the alloy is very dilute, so that the imperfections only become important in electron scattering at much lower temperatures, L^e can fall well below L_0 before temperatures are reached where the imperfections dominate and raise L^e back to L_0.

There are, then, two simple ways in which the electronic thermal conductivity of an alloy can be estimated from measurements of the electrical conductivity, and these will tend to produce the same results for impure alloys and for high and low temperatures. One way is to assume that the impurity concentration is sufficient so that L^e can never depart appreciably from L_0. The electronic thermal conductivity is then obtained from the electrical conductivity at the same temperature by direct application of the relation $\kappa^e/\sigma T = L_0$. This method must fail at 'intermediate' temperatures if the alloy is not impure enough and L^e falls below L_0. The other way is to assume that the electronic thermal resistivity can be considered as the sum of an ideal resistivity due to electron–phonon scattering and a resistivity due to imperfections. The latter is obtained from the residual electrical resistivity ρ_0 as $T \to 0$ K, by the relation $(W^e)_{\text{defects}} = \rho_0/L_0 T$. In the simplest case the ideal resistivity would be taken to be the same as that in the parent metal, which is little different from the measured thermal resistivity of a pure metal except at low temperatures. This method must fail if there is sufficient impurity in the alloy to alter either the electronic or phonon properties, or both, so that the resistance due to the electron–phonon interactions is different in the alloy from that in the pure metal.

These methods can be illustrated by predicting the electronic thermal conductivity of a Cu–10% Zn alloy which was measured by Kemp, Klemens, and Tainsh (1957). They determined $\rho_{90\text{K}} = 2 \cdot 31 \times 10^{-8}$ Ω-m and $\rho_0 = 1 \cdot 94 \times 10^{-8}$ Ω-m. For copper it is known from several analyses of the thermal conductivity of fairly pure specimens that, at 90 K, $W^e_p = 1 \cdot 8 \times 10^{-3}$ m K W^{-1}. The first method of estimating the electronic thermal conductivity of the alloy at 90 K then gives $\kappa^e = L_0 T/\rho_{90\text{K}} = 95$ W m^{-1} K^{-1}. The second method gives the electronic thermal resistivity

$$W^e = W^e_p + \rho_0/L_0 T = 10^{-3}(1 \cdot 8 + 8 \cdot 8) = 1 \cdot 06 \times 10^{-2} \text{ m K W}^{-1}$$

and $\kappa^e = 94$ W m^{-1} K^{-1}.

The two values are nearly equal to one another and are very similar to the measured thermal conductivity of 110 W m^{-1} K^{-1} at this temperature. If the exact numbers were taken seriously, one would deduce that the lattice conductivity is ~ 15 W m^{-1} K^{-1}, which is close to the estimate given in Table 12.1 for pure copper. This remarkable agreement is, however, somewhat fortuitous.

For a copper–16% Au alloy, similar calculations from the electrical resistivity give electronic thermal conductivities at 90 K equal to 29·8 and 29·5 W m^{-1} K^{-1} respectively. These are again very close to one another, but lead to a lower estimate for the lattice conductivity than in the zinc alloy, since the measured thermal conductivity was only 31 W m^{-1} K^{-1}.

As long as the computed electronic thermal conductivity is less than the measured conductivity, it is not immediately obvious whether the calculations are correct. The difference can just be ascribed to the lattice conductivity. For most practical purposes addition of the two main components of electronic thermal resistivity will provide sufficient accuracy. However, in experiments on dilute tin–cadmium alloys (with cadmium content below 1 per cent) Karamargin, Reynolds, Lipschultz, and Klemens (1972b) found that the behaviour of the lattice conductivity, deduced as the difference between the presumed electronic thermal conductivity and the total measured conductivity, was very complicated. This lattice conductivity increased with temperature initially from the lowest temperature of the measurements (4·2 K), but then decreased very rapidly above a temperature which was very different for each specimen. It thus had very different values just where the conductivity might be expected to be little dependent on impurity content, being largely determined by phonon–phonon interactions. They had previously (Karamargin, Reynolds, Lipschultz, and Klemens 1972a) determined the deviations from Matthiessen's rule in the electrical resistivity. At any temperature for each specimen they determined $\Delta \rho_i$, the amount by which the measured electrical resistivity was greater than the sum of the ideal resistivity, determined from measurements on pure tin, and the residual resistivity. They assumed that there was a similar deviation from additivity of the thermal resistivities, with the extra resistance given by $\Delta W_i^e = \Delta \rho_i / L_0 T$. When this additional thermal resistivity was included in W^e and the appropriate conductivity was subtracted from the total measured, the derived lattice conductivity did behave in a more orderly manner. Although such an effect is extremely important when trying to deduce values for lattice conductivity in dilute alloys, in the cases quoted the additional electronic thermal resistivity was at most only about 10 per cent of the resistivity itself.

12.2.2. *The lattice conductivity in metals and alloys*

As described above, the lattice thermal conductivity can be determined fairly accurately in an alloy, and by extrapolating this conductivity as a function of impurity content to zero impurity concentration, the lattice conductivity in a pure metal may be deduced. Having derived the lattice conductivity and its temperature variation it is of interest to compare it with theoretical predictions. In view of the variation of the lattice resistivity due to electrons with electron mean free path, it is simplest to compare theory and experiment first for pure metals, where l_e is largest and the results of perturbation theory should hold best.

In an ideal metal lattice thermal resistivity would be determined only by phonon–phonon U-processes and by electron–phonon interactions. Lindenfeld and Pennebaker (1962) have shown that dislocations in well-annealed copper alloys do not contribute appreciably to the lattice resistivity, and we may assume that this is a general result which will apply to metals which can be well annealed. The resistivity due to boundary scattering can be estimated in the same way as for dielectric crystals. In copper, for example, the boundary lattice resistivity is $\sim 1\cdot 5 \times 10^{-4}/l_B^p T^3$ m K W^{-1}, where l_B^p m is the effective mean free path for boundary scattering. For the boundary resistance to be comparable with the resistance due to electron–phonon scattering, $l_B T$ would have to be less than $\sim 2 \times 10^{-5}$ m K. This might suggest that boundary scattering is important below ~ 10 K for specimens composed of crystallites of the order of microns in size. However, most specimens have nearly single-crystal density and it is unlikely that phonon mean free paths would be confined to the crystallite size for wavelengths corresponding to the appropriate temperatures. We can, therefore, ignore boundary scattering for phonons in metals.

Although the direct effect of impurity scattering is removed by extrapolating results to zero impurity concentration, it is comforting to have some idea of the effect of impurities. Impurities are most important at intermediate temperatures, since U-processes dominate at high temperatures and electron–phonon interactions at low temperatures. Garber, Scott, and Blatt (1963) estimated that at 70 K the thermal resistivity due to 0·65 per cent of tin in copper was ~ 40 per cent of the resistivity due to U-processes. Below ~ 30 K electron–phonon scattering has become dominant. We may safely conclude that with purities easily achieved in metals the impurities cannot make an appreciable contribution to the lattice resistivity even in the temperature range between the regions where U-processes and electrons dominate the scattering.

The lattice conductivity of a metal at high temperatures can be compared with the resistivity due to U-processes calculated from eqn (7.3a), while the conductivity at low temperatures can be compared with expressions which can be derived from eqns (11.2), (11.3), and (11.4). From these equations we can express the lattice thermal conductivity in terms of (a) the ideal electronic thermal conductivity at the same temperature, (b) the ideal electrical conductivity at the same temperature, or (c) the electronic thermal conductivity in the limit of high temperatures. The relation between the low-temperature electronic thermal conductivity and the electrical resistivity, and between the low-temperature and high-temperature electronic thermal conductivities have already been discussed

(§ 12.1.2(a)), so that here the low-temperature lattice resistivity W_e^p will only be compared with the low-temperature electronic ideal thermal resistivity W_p^e which is due to the same interactions. If electrons interact with all phonons the relation between the two resistivities is given by

$$W_e^p = \frac{n_a^{\frac{4}{3}}}{351}\left(\frac{\theta}{T}\right)^4 W_p^e \qquad (12.2)$$

on the usual assumptions of spherical Fermi surface and Brillouin zone. If, however, electrons only interact with longitudinal phonons, W_p^e is reduced, but if phonon–phonon N-processes are very frequent all phonons will be affected by the electron scattering. In this extreme situation the relation between the two conductivities would be

$$W_e^p = 3\frac{n_a^{\frac{4}{3}}}{351}\left(\frac{\theta_L}{T}\right)^4 W_p^e \qquad (12.3)$$

where θ_L is the Debye temperature corresponding to longitudinal phonons.

We may consider a few experiments on alloys from which the lattice conductivity can be derived at both low and 'high' temperatures.

12.2.2(a). Silver. Kemp, Klemens, Sreedhar, and White (1956) measured the thermal conductivity of a series of annealed silver–palladium and silver–cadmium alloys. They assumed that the electronic thermal resistivity was made up of two components: an ideal resistivity which was the same as in pure silver and a resistance due to the impurities. The latter is inversely proportional to the temperature and is derived from the limiting low-temperature behaviour of the measured conductivity, where the ideal electronic thermal resistivity is negligible and the lattice conductivity makes a negligible contribution to the measured conductivity.

The calculated electronic thermal conductivity was subtracted from the measured conductivity and this difference was ascribed to lattice conduction. The lattice conductivity was nearly proportional to T^2 below ~10 K in all cases, and its absolute magnitude decreased gradually with increasing impurity. At these temperatures the electronic thermal conductivity is entirely determined by the impurities and decreases rapidly with increasing alloying. As a result, κ^p contributes about one-third of the measured conductivity of an alloy containing 2 per cent Pd at 10 K, but contributes about nine-tenths of the much smaller measured conductivity of an alloy with 40 per cent Pd. At 1 K the electronic conductivity has only decreased 10-fold, being proportional to the temperature, while the lattice component is 100 times less than at 10 K, so that the two components are about equal at 1 K even for the 40 per cent alloy.

Extrapolation of the lattice resistivity to zero concentration of impurities should give the lattice resistivity in the pure metal. Extrapolation for both the palladium and cadmium alloys leads to a lattice resistivity at low temperatures given by $W^P T^2 \sim 5 \text{ m K}^3 \text{ W}^{-1}$. This cannot be compared directly with the ideal electronic thermal resistivity because the latter was found to be proportional to $T^{2.4}$ rather than to T^2. However, we may make a comparison of the magnitudes at 10 K, where the 'observed' value of W^P is 5×10^{-2}, while the value derived from W_p^e by eqn (12.2) is $3 \times 10^{-2} \text{ m K W}^{-1}$, if it is assumed that electrons interact equally with phonons of all polarizations. This assumption receives some support from the relatively good agreement in the values. The other extreme assumption represented by eqn (12.3) would lead to a value of W_e^p about 20 times higher (apart from the visible factor of 3, θ_L^4 is many times greater than θ^4 derived from specific-heat measurements).

For small concentrations of both palladium and cadmium the lattice conductivities are similar at 100 K at a value of just over $10 \text{ W m}^{-1} \text{ K}^{-1}$ and are decreasing with increasing temperature. Eqn (7.3a) gives a high-temperature lattice conductivity $\kappa_p^P \sim 900/T \text{ W m}^{-1} \text{ K}^{-1}$, and again agreement with the theory is quite satisfactory.

At 100 K the lattice contributes about one-twentieth of the total in the 2 per cent Pd alloy, but over one-quarter in the 40 per cent Pd alloy. These proportions decrease with increasing temperature because the lattice conductivity is decreasing while the electronic contribution is increasing.

12.2.2(b). Copper. Similar experiments on annealed copper–zinc alloys by Kemp *et al.* (1957) gave a limiting lattice resistivity at low temperatures for zero zinc concentration $W^P T^2 \sim 5 \cdot 5 \text{ m K}^3 \text{ W}^{-1}$. As for silver, the ideal electronic thermal resistivity W_p^e was found to be proportional to $T^{2.4}$, so that the comparison has to be made at specific temperatures. At 10 K the 'observed' value of W^P is $5 \cdot 5 \times 10^{-2}$, while that derived from W_p^e is between 4 and $8 \times 10^{-2} \text{ m K W}^{-1}$ (see Table 12.1).

At 100 K the lattice conductivity in the most dilute alloys is a little over $10 \text{ W m}^{-1} \text{ K}^{-1}$, while eqns (7.3a), (11.3), and (11.4) give a conductivity resulting from a combination of phonon–phonon U-processes and electron–phonon scattering of $15 \text{ W m}^{-1} \text{ K}^{-1}$.

12.2.2(c). Tin. Karamargin *et al.* (1972b) analysed their measurements of the thermal conductivity of tin–cadmium alloys. The lattice resistivity they deduce for the lowest temperatures is given by $W^P T^2 = 70 \text{ m K}^3 \text{ W}^{-1}$. By using a value of 145 K for θ, corresponding to an average temperature for the measurements (the limiting value θ_0 is 195 K), and taking n_a as 4, they derive from W_p^e a value for $W_e^p T^2 = 80 \text{ m K}^3 \text{ W}^{-1}$.

Karamargin *et al.* showed that for the three most concentrated alloys the lattice conductivity at the highest temperature of the measurements (70 K) is appreciably influenced by scattering by electrons, impurities, and the tin isotopes. At 70 K the lattice conductivities are not very different among the different alloys and for a cadmium content of 0·74 per cent the value is ~5 W m^{-1} K^{-1}. Eqn (7.3a) suggests that, if only U-processes were important and the resistivity due to them were proportional to T, the conductivity would be $500/T$, so that we would not expect it to be greater than 7 W m^{-1} K^{-1} at 70 K, which accords well with the measurements.

12.2.2(d). Lattice conductivity at 'high' temperatures. At temperatures such that the electronic Lorenz number would be expected to approach the ideal value L_0 the lattice conductivity can be estimated from the amount by which the measured Lorenz number $\kappa/\sigma T$ exceeds L_0, if it is assumed that the presence of a lattice component is the only reason for L to depart from L_0. For example, Moore, McElroy, and Barisoni (1966) determined values for L of 2·59 and 2·61 × 10^{-8} W Ω K^{-2} on two different specimens of platinum at 300 K. From the plot of L^e as a function of T/θ for pure metals given by Hust and Sparks (1973) (Fig. 12.2), it can be seen that L^e for platinum at 300 K (~1·3θ) is slightly less than 2·4 × 10^{-8}. If the observed L is assumed to exceed this value only because the lattice contributes to the total thermal conductivity used in deriving L, then κ^p is 0·2/2·6 of the observed conductivity, which was 70·5 and 72·1 W m^{-1} K^{-1} for the two specimens. On this assumption $\kappa^p \sim 5\cdot 3$ W m^{-1} K^{-1}. From eqn (7.3a), taking $\nu = 4$ and $\gamma = 2\cdot 5$, the lattice conductivity of platinum limited only by phonon–phonon U-processes would be $1200/T$ and thus ~4 W m^{-1} K^{-1} at 300 K.

This agreement is satisfactory in view of the difficulty of predicting the lattice conductivity resulting from U-processes in terms of other parameters of a crystal. However, utilizing departures from L^e cannot be regarded as a universal method for deriving lattice conductivities of pure metals, since for relatively small lattice conductivities such departures cannot be deduced with sufficient accuracy, and for complex metals it is not established that κ^e is always given by applying L^e, taken from Hust and Sparks, to the measured electrical resistivity.

12.2.3. The dependence of lattice conductivity on electron mean free path. There have been several sets of experiments to test the applicability of the Pippard theory. Since the electronic thermal conductivity must be subtracted from the total measured, it is simplest to work at helium temperatures where the electronic thermal conductivity can be deduced from the electrical conductivity with some confidence in the validity of the WFL law. The electron mean free path is directly related to the residual electrical resistivity ρ_0.

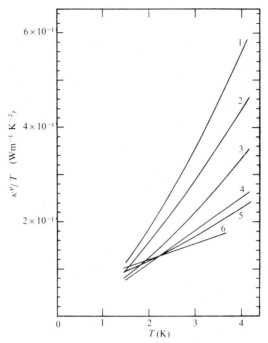

FIG. 12.3. The lattice conductivity, plotted as κ^p/T as a function of temperature for copper–germanium alloys. The residual electrical resistivities increase from specimen 1 to specimen 6, the ratio ρ_0^6/ρ_0^1 being ~40. (After Lindenfeld and Pennebaker 1962.)

Zimmerman's (1959) measurements were on silver–antimony alloys with ρ_0 from 1·24 to 3·06 × 10^{-7} Ω m, while those of Lindenfeld and Pennebaker (1962) were on copper alloys with germanium (six alloys), aluminium, gallium, or indium (one alloy each) with ρ_0 from 0·03 to 1·29 × 10^{-7} Ω m. The latter authors show how results can be scaled for different alloy systems and include Zimmerman's results in their discussion.

Fig. 12.3 shows κ^p/T plotted against T for the six alloys with germanium measured by Lindenfeld and Pennebaker. It is immediately apparent that the lattice conductivities for all the alloys are very different and range over a factor of ~3 at 4 K as ρ_0 varies by a factor of 40. In addition, the curves do not even approximate to straight lines through the origin, which would correspond to $\kappa^p \propto T^2$. From the theory mentioned in § 11.3.3 it follows that all the results for a single solvent metal should fall on a universal curve if $\kappa^p/T\rho_0$ is plotted against T/ρ_0. Fig. 12.4 shows such a plot for the germanium alloys (the Al, Ga, and In alloys fall in the appropriate positions, while Zimmerman's results for T/ρ_0 between ~7 × 10^6 and 5 × 10^7 K Ω^{-1} m^{-1}, when suitably scaled for the different solvents, fall closer to the theoretical curve). The theoretical curve shown was

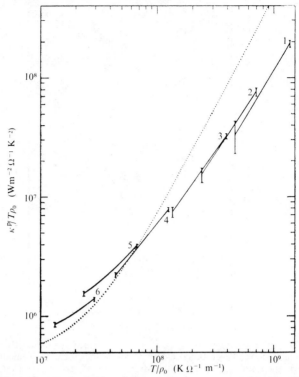

FIG. 12.4. The lattice conductivities of the specimens represented in Fig. 12.3, expressed as $\kappa^p/T\rho_0$, as a function of electron mean free path, expressed as T/ρ_0. The dotted line was calculated from the mean free paths illustrated in Fig. 11.4. (After Lindenfeld and Pennebaker 1962.)

computed on the assumption that the transverse and longitudinal modes make independent contributions to the thermal conductivity.

The slope of the theoretical curve at high T/ρ_0 on the logarithmic plot approaches 2, because of the assumed independent behaviour of transverse phonons. This would imply a lattice conductivity proportional to T^3, whereas the experimental slope is nearer to linear here, corresponding to $\kappa^p \propto T^2$ and independent of ρ_0 for very small ρ_0. For low T/ρ_0, $\kappa^p/T\rho_0$ approaches a constant value, corresponding to $\kappa^p \propto T\rho_0$.

Zimmerman's measurements span values of T/ρ_0 between the two limiting regimes, and he was able to express the conductivity of each alloy in the form $\kappa^p = aT + bT^2$, where a increases slightly between the smallest and largest value of ρ_0, while b decreases steadily with increasing ρ_0.

Tainsh and White (1962) measured a series of copper alloys and a series of silver alloys. They found that below 10 K the lattice conductivity was proportional to T^2 but varied with the solute concentration. They did not, however, give a detailed analysis in terms of the Pippard theory.

Garber et al. (1963) measured the thermal conductivity of three copper–tin alloys with ρ_0 between 1·75 and $4·86 \times 10^{-8}$ Ω m, and of various ternary alloys of copper with ρ_0 between 0·55 and $1·06 \times 10^{-8}$ Ω m between 10 and 90 K. The derivation of the lattice conductivity was slightly less certain than in experiments at helium temperatures, because it was necessary to assume that the electronic component was given by adding the ideal resistivity W_p^e to the impurity resistivity given by $\rho_0/L_0 T$. However, the lattice conductivities deduced in this way were sufficiently different from one another for the main features revealed not to be affected by slight departures from simple additivity for the electronic component. Below ~20 K the lattice conductivities were proportional to T^2 and decreased with increasing ρ_0. For the temperature range of the measurements, the alloys used did not provide low enough values of $T/\rho_0 (<5 \times 10^7$ K Ω^{-1} m^{-1}) for marked departures from the T^2 relation to be observed, as in the experiments of Zimmerman and of Lindenfeld and Pennebaker for alloys of high residual resistivity at helium temperatures.

Lindenfeld and Pennebaker and also Garber et al. plotted graphs of $W^p T^2$ for various alloys (mainly of copper) measured by a number of authors. These show that in the limit of $\rho_0 \to 0$, $W^p T^2$ for copper is about 5 m K^3 W^{-1}, in agreement with the value quoted in § 12.2.2(b), but rises to five times this value for ρ_0 about 10^{-7} Ω m. At the opposite extreme, the alloys measured by Zimmerman had such high residual resistances that at helium temperatures T/ρ_0 was so small that in his expression $\kappa^p = aT + bT^2$ the term in T^2 was never dominant, so that the departure from a T^2 dependence was correspondingly great.

12.2.4. *The influence of dislocations on the lattice conductivity of alloys.*

The effect of dislocations on electronic thermal conductivity can be deduced from their effect on electrical conductivity. In order to observe their influence on lattice conductivity, it is necessary to be able to deduce the lattice conductivity with considerable accuracy from the measured conductivity. For this reason investigations aimed at studying the relation between the number and nature of dislocations and the corresponding lattice conductivity have been carried out on alloys and on superconductors.

The main term which arises in the scattering rate for phonons by sessile dislocations leads to a lattice resistivity proportional to T^{-2}, which is the same behaviour as results from scattering of phonons by electrons if T/ρ_0 is not too small. Direct interpretation of experimental results thus depends on the deformation and annealing carried out in the investigation not changing the strength of the electron–phonon interaction or changing

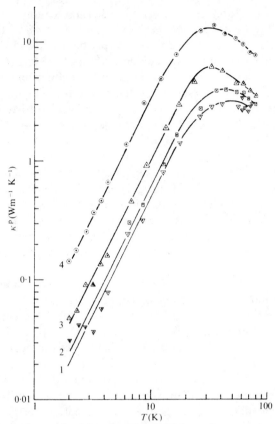

FIG. 12.5. The lattice conductivity of torsionally deformed copper–zinc specimens: 1, as deformed; 2, annealed up to 250 °C; 3, annealed up to 290 °C; 4, annealed up to 400 °C. During annealing the temperature was increased by 6 K per minute. (After Kemp *et al.* 1959.)

ρ_0 too much. If these criteria are satisfied, the two components of resistivity which both vary as T^{-2} can be distinguished from one another.

Some typical experimental results are shown in Fig. 12.5, taken from the work of Kemp, Klemens, and Tainsh (1959) on torsional deformation of a copper–zinc alloy. After deformation, the lattice conductivity is proportional to T^2 at temperatures below ~20 K and recovers towards its initial values on annealing. In these experiments the dislocation density was estimated from measurements of the release of stored energy during annealing of specimens made from the same material. In the experiments of Lomer and Rosenberg (1959), also on Cu–Zn alloys, comparison was made between changes in lattice conductivity and the dislocation densities measured on electron micrographs obtained from thin sections of the material.

In these and in several subsequent investigations the number of dislocations deduced from thermal conductivity, using eqn (8.3) for the scattering by the strain field of a screw dislocation, exceeds estimates from the various kinds of auxiliary experiments by a factor of about 10. This implies that the scattering by dislocations is stronger than is predicted. The discrepancy is at least an order of magnitude less than is observed for several pure dielectric crystals, and if the large discrepancy in these cases is due to the larger scattering by vibrating dislocations (see § 8.3.4), then perhaps in many alloys this kind of scattering is much less important or even negligible.

Ackerman (1972) has made several suggestions as to how the discrepancy might be decreased. The expression for the scattering which is generally used is that for a screw dislocation perpendicular to the temperature gradient, multiplied by a suitable averaging factor to represent a random array of dislocations. Following Schoek (1962), Ackerman suggests another averaging procedure which takes into account the actual total length of dislocation lines passing through a volume when they are arranged randomly. This increases the averaging factor by almost 3. He also shows that the scattering by an edge dislocation for the same magnitude of Burgers vector is $\frac{13}{8}$ times that of a screw dislocation. If the Burgers vector is randomly oriented with respect to the dislocation line, there will be twice as many edge dislocations as screw dislocations, so that the total scattering is 1·4 times greater than if all dislocations were screw. Putting both these effects together, the discrepancy is reduced to the order of 2, without considering the accuracy of deducing the number of dislocations from other measurements.

Detailed studies of the lattice component of the conductivity of deformed alloys have revealed deviations from a simple temperature dependence of the resistivity due to dislocations. Leaver and Charsley (1971) observed a change in slope of the lattice conductivity of deformed copper alloys. They suggested that the overall long-range strain field of a close arrangement of dislocations is less than the strain field of a random arrangement of individual dislocations. This reduction is felt by phonons with wavelengths greater than the separation of dislocations, and is reflected in a change in the temperature dependence of the conductivity at a temperature corresponding to the dominance of these phonons. Kapoor, Rowlands, and Woods (1974) observed changes of slope in plots of κ^p/T versus T for cold-worked noble-metal alloys. They pointed out that although the temperatures at which the kinks occurred for the three metals differ, the dominant wavelengths at these temperatures are all about the same. From a study of the temperature variation of the lattice resistivity of deformed copper–aluminium alloys, Mitchell, Klemens, and Reynolds (1971) deduced that in the impurity atmosphere around a dislocation an aluminium atom occupied a volume 23 per cent larger than a copper atom.

Evidence that phonon scattering by dislocations in metals is not necessarily due to sessile dislocation comes from work on superconductors. The thermal conductivity of a superconductor well below the transition temperature is entirely due to phonons (see the next section). Anderson and his co-workers (Anderson and Smith 1973, O'Hara and Anderson 1974a, b) have examined the influence of dislocations on the thermal conductivity of superconducting niobium, aluminium, lead, and tantalum down to about 0·04 K. In all cases the phonon scattering was much greater (up to $\sim 10^4$ times greater) than could be due to sessile dislocations, and they ascribed the observations to resonant scattering due to vibrating dislocations. For lead and tantalum the phonon mean free path due to the dislocations passed through a minimum which shifts in temperature with the strain, while in aluminium and niobium the minimum does not shift. It was concluded that in the first two metals the vibrating dislocations may be represented by the elastic-string model (Garber and Granato 1970), while for the second two metals a better description is of the dislocations vibrating in the Peierls potential.

12.3. Superconductors

When a metal or alloy becomes electrically superconducting, the electrons which take part in the condensation process are unable to transport energy and to interact with phonons. If the phonon component of the thermal conductivity is negligible in the normal state, then below the transition temperature the thermal conductivity of the metal in the superconducting state is lower than that in the normal state. (The normal-state values are established either by extrapolation from higher temperatures or by measurement in magnetic fields which are sufficient to destroy superconductivity.) This occurs because the reduction in the effective number of electrons is dominant in controlling the electronic thermal conductivity.

The simplest type of behaviour is shown by pure metallic type-I superconductors in which phonon conduction can be neglected down to temperatures well below the transition temperature T_c. If T_c lies below the temperature of the thermal-conductivity maximum, the metal becomes superconducting at a temperature such that the electron mean free path in the normal state is almost entirely limited by defect scattering and is thus independent of temperature. If we make the assumptions that in the superconducting state the mean free path for all the effective electrons remains at this constant value and that the effective electron velocity is unchanged, then the ratio of superconducting- to normal-state thermal conductivities should be equal to the ratio of the corresponding specific heats. The expression for the electronic specific heat of a superconductor, given by the Bardeen–Cooper–Schrieffer (BCS) theory (Bardeen, Cooper, and Schrieffer 1957) is somewhat complicated, but for $T < \sim 0.4 T_c$ it reduces to an exponential temperature dependence. The ratio $\kappa^e(s)/\kappa^e(n)$ of the electronic thermal conductivities would, on our simple assumptions, then be proportional to $\{\exp(-a/T)\}/T$ at low temperatures, where a is a constant.

Bardeen, Rickayzen, and Tewordt (BRT) (1959) have given a theory of the thermal conductivities of superconductors, based on the BCS theory, and showed that the ratio $\kappa^e(s)/\kappa^e(n)$ is unity at T_c and then follows an exponential form quite well as the temperature decreases. Although the low-temperature behaviour predicted by the BRT theory is in accord with that expected from our simple assumptions, this is not so near T_c, where there is a discontinuity in the specific heat (with $C^e(s)/C^e(n) = 2.4$ according to the BCS theory), but the thermal-conductivity curve for the superconducting state falls away from the normal-state curve without a discontinuity.

Zavaritskii (1957) measured the thermal conductivity of specimens of tin which were of such purity that $T_c(3.7 \text{ K})$ was just below the temperature of the conductivity maximum, as can be seen in Fig. 12.6a. From $T_c/T = 1$ down to $T_c/T \sim 10$ the conductivity could be represented by $\kappa(s) \propto \exp(-1.45 T_c/T)$. The difference between the behaviours of the specific heat and of the thermal conductivity in the superconducting state at T_c showed up in the measured thermal diffusivity (k/C), which decreased discontinuously when the metal became superconducting at T_c as shown in Fig. 12.6b.

Zavaritskii found that below 0·4–0·6 K, depending on the specimen (quality, dimensions, and surface finish were all relevant), the thermal conductivity quite sharply ceased to fall as rapidly as the exponential and became proportional to a relatively small power of the temperature (see Fig. 12.6a). This conductivity corresponded to the lattice conductivity to be expected for a phonon mean free path of the same order of magnitude as the specimen diameter. The electronic thermal conductivity is then too low to be an appreciable contribution, and, in addition, the scattering of phonons by electrons is negligible.

The same kind of behaviour was found (Zavaritskii 1958) for aluminium ($T_c = 1.2$ K) and for zinc ($T_c = 0.84$ K). In the former the exponent in the thermal-conductivity variation was $-1.5 T_c/T$ and lattice conductivity became dominant at ~ 0.2 K, while for the latter the exponent was $-1.3 T_c/T$ but there was no definite indication of the lattice conductivity down to ~ 0.14 K.

If T_c lies above the temperature of the thermal-conductivity maximum, the electronic component is decreased by the reduction in the effective

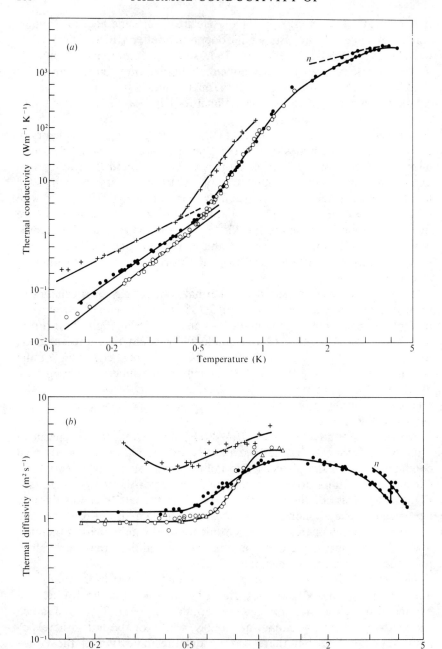

FIG. 12.6. The thermal conductivity (a) and thermal diffusivity (b) of three tin specimens in the superconducting state. The curve for the normal state is marked n. (After Zavaritskii 1957.)

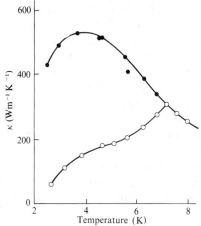

FIG. 12.7. The thermal conductivity of pure lead: ● in a magnetic field to keep the metal normal below the transition temperature; ○ no magnetic field and thus superconducting below the transition temperature. (After Olsen 1952.)

number of electrons but is increased by the reduction in phonon–electron scattering. As a result, while breaking away sharply from the rising $\kappa^e(n)$, $\kappa^e(s)$ does not fall off rapidly with decreasing temperature until well below T_c, when any further reduction in phonon–electron scattering cannot enhance $\kappa^e(s)$ any further because defect scattering has become dominant. This is shown for lead in Fig. 12.7.

The lattice conductivity in a metal or alloy at low temperatures is usually limited by phonon–electron interactions, so that κ^p increases below T_c when the metal or alloy becomes superconducting. Although this rise in lattice conductivity is generally small compared with the fall in electronic thermal conductivity, the opposite situation can occur in sufficiently disordered alloys (for which both κ^p and κ^e are small) in which phonons provide a large fraction of the normal-state thermal conductivity. The thermal conductivity is then higher in the superconducting state than in the normal state, as shown in Fig. 12.8.

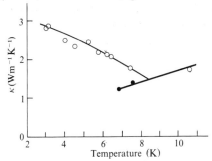

FIG. 12.8. The thermal conductivity of lead–30% bismuth: ● normal below the transition temperature; ○ superconducting below the transition temperature. (After Olsen 1952.)

The behaviour of superconductors in magnetic fields can be very complex and will not be discussed here (a discussion and a number of references can be found in Saint-James, Sarma, and Thomas 1969). Type-I superconductors in the intermediate state (which occurs for fields between the critical field, defined for a long rod lying parallel to the field, and some fraction of it, which depends on the shape and orientation of the specimen) consist of superconducting and normal lamellae; apart from the different thermal conductivities of these two states as bulk material, there are effects due to the boundaries between them (see, for example, Mendelssohn and Schiffman 1960). Type-II superconductors are even more complicated owing to the existence of the 'vortex state' for a range of magnetic fields between the critical fields H_{c1} (when magnetic flux starts to penetrate the specimen) and H_{c2} (at which normal electrical resistivity reappears).

For pure metals the critical magnetic field required to restore a superconductor to the normal state can be quite low, even near 0 K, so that the thermal conductivity can easily be switched from the normal to the superconducting value. The ratio $\kappa(s)/\kappa(n)$ decreases almost exponentially with decreasing temperature while $\kappa(s)$ is dominated by the electronic component, but this ratio continues to decrease (as T^3/T) when $\kappa(s)$ is dominated by the phonon component. A thermal switch can thus be made which requires only a small magnetic field for its operation. The first such switch, made of tantalum, was operated by Heer and Daunt (1949) and at 0·7 K the ratio of the heat flows in the closed (normal) and open (superconducting) states was ~1000. Reese and Steyert (1962) found that for polycrystalline lead between 0·1 and 0·5 K $\kappa(s)/\kappa(n) \sim 0·02 T^{2·3}$, the numerical constant being dependent on the orientation of the magnetic field.

Switches of this kind would have been an essential component of cyclic demagnetization refrigerators for obtaining temperatures below 0·3 K (attainable by pumping on liquid ^{3}He). Although superconducting switches are still commonly used, this particular application did not come to fruition because the method of cooling was superseded by the advent of the ^{3}He/^{4}He dilution refrigerator.

13

SEMICONDUCTORS

THERE are several features of thermal conduction in semiconductors which warrant separate treatment. In a pure semiconductor the conductivity at normal and low temperatures may be due entirely to the lattice, so that its behaviour is similar to that found in non-metals, which has already been described. The introduction of small amounts of impurity first produces a reduction in the phonon conductivity due to scattering by the impurity ions themselves and, in many cases, by the electrons assiciated with their presence. The latter can have different characteristics from the scattering by electrons forming a degenerate system (where only the electrons near the Fermi energy are able to engage in scattering). With sufficient doping the electronic thermal conductivity becomes appreciable, but, if the electron system is still not degenerate, the relation between electrical conductivity and electronic thermal conductivity is not the same as in ordinary metals. There is an additional means by which electrons and holes can conduct heat in a semiconductor. Electron-hole pairs are created at the hot end and drift down the temperature gradient to recombine at the cold end. The ionization energy is transmitted through the semiconductor.

The thermal conductivity of semiconductors is a relevant property in considerations of thermoelectric refrigeration. Although the 'figure of merit' of a device depends on the combined properties of a pair of materials which are used together, the general requirement is a large thermoelectric power, a high electrical conductivity, and a low thermal conductivity. The rate of removing heat depends on the thermoelectric power, but the unwanted heat generated by the current which produces cooling by the Peltier effect depends on the electrical conductivity. The adverse conduction of heat to the cooled body depends on the thermal conductivities of the semiconductors forming the junctions. The reduction in the thermal conductivity of silicon produced by the addition of germanium has been mentioned in § 8.3.1(b).

13.1. Pure intrinsic semiconductors

The general behaviour of the thermal conductivity of pure intrinsic semiconductors can be illustrated by considering results obtained for pure silicon and germanium. Although the results of different authors differ slightly, the agreement is sufficient for us to have confidence in the magnitude of the conductivity and in the form of its temperature dependence. The work of Glassbrenner and Slack (1964) will be described here because they measured the conductivities over a wide temperature range: from 3 K to 1580 K for silicon and to 1190 K for germanium, the upper temperature in each case being slightly below the melting point. Their results, obtained by a radial heat-flow method, agree quite well with those obtained by Abeles *et al.* (1962) and Beers, Cody, and Abeles (1962) who measured the thermal diffusivity. There is also agreement with the radial heat-flow measurements of Fulkerson, Moore, Williams, Graves, and McElroy (1968) on silicon up to 1300 K.

The conductivities below about room temperature can be explained in terms of scattering by three-phonon U-processes, boundaries and isotopes, and since there are appreciable concentrations of isotopes in both elements in nature the analysis can be made without invoking N-processes.

The thermal resistivity above room temperature due to isotopes, W_I, should be almost independent of temperature (see § 8.2.1), and, when this is subtracted from the measured thermal resistivity, we would expect the remaining resistance to be proportional to temperature, apart from any effects due to thermal expansion (see § 7.1.1(a)). Glassbrenner and Slack plotted $(W - W_I)/T$ against T and found that this was by no means constant, as shown in Fig. 13.1. Slack (1972a) showed that the effect of thermal expansion is very small in these materials, and at 300 K it should only increase the power of T by 0·02 for germanium. He analysed W_p^p into a part proportional to T and a part proportional to T^2, the latter being ascribed to four-phonon processes as predicted by Pomeranchuk (1941). At the temperature $T = \theta$ the term in T^2 contributes one-third as much as the linear term to the resistivity of germanium and two-thirds as much for silicon.

A variation of conductivity faster than T^{-1} could also be due to the change in phonon frequencies with temperature, as suggested by Ranninger (1965), or to the effect of four-phonon processes on the contribution to the conductivity from low-frequency longitudinal phonons (Klemens and Ecsedy 1976) mentioned in § 7.1.1(a). While these explanations are quite general, Logachev and Vasil'ev (1973) suggested that in germanium, silicon, and III–V compounds, three-phonon processes involving both acoustic and optic phonons are important and are 'de-activated' more rapidly with decreasing temperature than processes involving only acoustic phonons. As acoustic phonons carry most of the heat, the temperature dependence of conductivity is faster than T^{-1}.

Above $\sim 1\cdot 6\theta$ the conductivities start to decrease less rapidly and $(W - W_I)/T$ reaches a maximum and then decreases again. The heat flow through the specimens by radiation was calculated and was shown not to

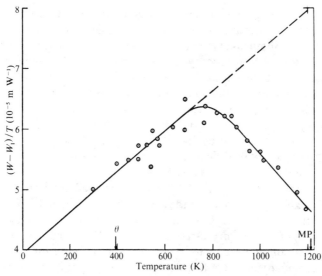

FIG. 13.1. The temperature dependence of $(W-W_\mathrm{I})/T$ for pure germanium; W is the measured thermal resistivity and W_I the calculated resistivity due to isotopes. The difference between the 'measured' curve and the extrapolated straight line above ~ 700 K is ascribed to electronic heat conduction. The calculated isotope resistivity is only 8 per cent of the total resistivity at 400 K and this fraction decreases with increasing temperature. Uncertainties in its estimation do not, therefore, influence conclusions about the additional mechanism acting above 700 K. (After Glassbrenner and Slack 1964.)

make an appreciable contribution to the conductivity measured, and the additional conductivity was ascribed to electrons. The electronic contribution consists of two parts, the energy flow associated with electrons and holes as for electrons in a metal, and the transport of ionization energy by electron–hole pairs.

If the electron and hole concentrations are low, then Boltzmann statistics apply. By taking proper averages of relaxation times over this distribution it is found that the Lorenz number associated with the transport of thermal energy is $L^\mathrm{e} = (k_\mathrm{B}/e)^2(\frac{5}{2}+p)$, where p represents the energy dependence of the relaxation time $\tau = \tau_0 E^p$. For scattering of electrons by acoustic-mode phonons $p = -\frac{1}{2}$, so that $L^\mathrm{e} = 2(k_\mathrm{B}/e)^2$ (see, for example, Blatt 1968). In this extreme of classical statistics this Lorenz number can be applied to the total electrical conductivity, whether it is due to electrons or to holes, to give the corresponding thermal conductivity.

From the measured electrical conductivity of silicon and germanium it is found that the corresponding thermal conductivity represents only a slight contribution to the conductivity of either of the pure materials up to the melting point (where it contributes 5 per cent of the total in silicon and 9 per cent in germanium).

Bipolar diffusion, which is the name given to the transport of ionization energy, has been discussed by Drabble and Goldsmid (1961), following Davydov and Shmushkevitch (1940) and Price (1955). For an intrinsic semiconductor with an energy gap which can be described by a single value E_g^T at a temperature T, the creation of an electron in the conduction band and a hole in the valence band extracts an amount of energy E_g from the high-temperature end of the specimen, and on recombination this energy is given up to the cold end. This transport of energy is additional to that carried by the electron and the hole separately, but is related to it. Bipolar diffusion is not an efficient process unless both electrons and holes have reasonable mobilities; the ratio μ' of the mobilities enters the expression for the conductivity symmetrically, so that for a given 'ordinary' conductivity the maximum bipolar diffusion conductivity occurs when both mobilities are equal. If, for example, the electronic mobility is very large and that of the holes is very small, the 'ordinary' conductivity due to electrons may be high, but the holes will not be accompanying them fast enough to provide the necessary numbers for recombination. The relation between the two conductivities also depends on the relation between relaxation time and energy; if $\tau \propto E^{-\frac{1}{2}}$ for both electrons and holes, and Boltzmann statistics are appropriate, the conductivities are related by the expression

$$\kappa_{bp} = \frac{\mu'}{(1+\mu')^2}\left(\frac{E_g^T}{k_B T}+4\right)^2 \frac{\kappa_t}{2}$$

where κ_{bp} is the conductivity due to bipolar diffusion and κ_t is the combined heat conductivity of electrons and holes conveying their thermal energy independently of one another. The ratio of electron and hole mobilities is given by μ'.

If the mobilities are nearly equal, then, for $k_B T \ll E_g$, κ_{bp} is much larger than κ_t, but since the number of electrons and holes is small both conductivities are very small. As the temperature is increased, the ratio κ_{bp}/κ_t decreases, and on the simple model described has a limiting value of 2.

Glassbrenner and Slack used published values and extrapolations for μ' (which is temperature dependent in Ge) and for E_g (which decreases considerably with increasing temperature in both Si and Ge). For Si at 1000 K and for Ge at 600 K, $\kappa_{bp} \sim 20\kappa_t$ but is still only ~ 2 per cent of the measured conductivity in both cases. At the melting points $\kappa_{bp} \sim 6\kappa_t$ for Si and $\sim 3\cdot 3\kappa_t$ for Ge and contributes 32 per cent of the total conductivity in both semiconductors. The calculation can be turned round to estimate the band gap and its temperature variation. Within the accuracy with which the gap can be determined in this way ($\sim 0\cdot 1$ eV), the authors find agreement with the extrapolations which were made from published values for lower temperatures.

Klein, Shanks, and Danielson (1963) measured the thermal conductivity of silicon up to 1400 K. They deduced a combined contribution from both kinds of electronic conduction which was 15–20 per cent higher than that found by Glassbrenner and Slack. This difference does not affect the general picture of the conductivity.

Martin, Shanks, and Danielson (1968) found that the thermal conductivity of Mg_2Sn was inversely proportional to temperature above 175 K, but decreased more slowly above room temperature. They calculated the electronic thermal and bipolar contributions from the known electrical properties and the band gap from 300 to 700 K. This contribution agreed quite well with the difference between the observed conductivity and an extrapolation of the lattice conductivity above 300 K. At 700 K the electronic component is about half the total, and at this temperature $\kappa_{bp} \sim 5\kappa_t$.

In InAs, measured by Stuckes (1960), the electron mobility is about 64 times greater than the hole mobility, so that $\kappa_{bp} < \kappa_t$. At 900 K $\kappa_{bp} \sim \frac{1}{3}\kappa_t$, and together they contribute about half the measured conductivity.

The thermal conductivities of many other pure semiconductors have been measured over wide temperature ranges. Results for III–V compounds above room temperature are given by Steigmeier (1969). A common feature is a variation of conductivity with temperature that is faster than $1/T$, the power of T being about $-1·2$ to $-1·3$ at room temperature and becoming more negative with increasing temperature. As for silicon and germanium, little of this extra temperature dependence can be ascribed to the change of volume with temperature. Measurements by Holland (1964), however, suggest that at $T \sim \theta$ the departures from the $1/T$ law are small for InSb and GaAs ($\kappa \propto T^{-1·1}$) and for GaSb ($\kappa \propto T^{-1}$) and the II–VI semiconductors CdS and CdTe ($\kappa \propto T^{-1}$).

The high-temperature conductivities of all the III–V compounds for which Steigmeier shows measured conductivities, as well as silicon and germanium, are included in Slack's (1977) discussion of the absolute magnitudes to be expected at temperatures where U-processes should lead to a $1/T$ behaviour (κ_{calc} derived from eqn (7.3)). The ratio $\kappa_{exp}/\kappa_{calc}$ at the Debye temperature appropriate to acoustic phonons, $\bar{\theta}$, lies between 0·7 and 0·9 in all cases but two, with GaSb giving a ratio 0·6 and GaP 1·1. As can be seen from Fig. 13.2 no obvious pattern emerges from the values of $\kappa_{exp}/\kappa_{calc}$ plotted against the mass ratio of the constituent atoms, contrary to what would be expected from the theory first proposed by Blackman (1935) and given some support from measurements on alkali halides (see Fig. 7.5). Similarly, no trend is shown by the II–VI compounds discussed by Slack (1972b), some of which he measured himself.

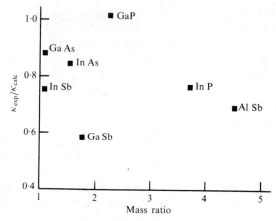

FIG. 13.2. The ratio of measured to calculated thermal conductivity of III–V compounds at θ, the Debye temperature corresponding to acoustic phonons, as a function of the ratio of masses of the constituent atoms. (Data from Slack 1977.)

13.2. Impure semiconductors

Impurities in a semiconductor can scatter electrons and phonons in the same ways as in metals and in non-metallic crystals (see, for example, Drabble and Goldsmid 1961). Impurities can also introduce free electrons and holes, and these can reduce the phonon conductivity by increasing the scattering. If the concentration of holes or electrons is great enough, they can contribute an appreciable component to the total thermal conductivity.

13.2.1. Semiconductor alloys

If the electrical properties are unchanged, a semiconductor has a better figure of merit for thermoelectric refrigeration if its thermal conductivity is reduced. Ioffe and Ioffe (1955) had suggested that appropriate impurities could lower the ratio of phonon conductivity to electron mobility, and Ioffe, Airapetyants, Ioffe, Kolomoets, and Stil'bans (1956) proposed that alloys of isomorphous semiconductors should thus have better figures of merit than the pure components.

Klemens (1960) gave a simple method for calculating the change in conductivity at high temperatures on introducing point defects. The effect of N-processes was ignored and only point defects, scattering phonons as ω^4, and U-processes, assumed to scatter as $T\omega^2$, were taken into account. The ratio of the conductivity of the alloy to that of the pure crystal at the same temperature can then be expressed as

$$\frac{\kappa_a}{\kappa_U} = \left(\frac{\omega_0}{\omega_D}\right) \tan^{-1}\left(\frac{\omega_D}{\omega_0}\right)$$

where ω_D is the Debye maximum frequency and ω_0 is the frequency at which the relaxation rates for scattering by the impurities and by U-processes are equal at the temperature concerned. If scattering by the impurities is assumed to be entirely due to the mass difference, the coefficient of the ω^4 term can be calculated (see § 8.1.1); ω_0 can then be derived from this coefficient together with the coefficient of the U-process scattering term derived from the measured conductivity of the pure crystal. A plot of the measured κ_a/κ_U as a function of the calculated ω_D/ω_0 for a number of alloys which have been measured by various authors is given by Drabble and Goldsmid (1961) and shows remarkable agreement with the simple expression above given by Klemens. For two of the alloy series the measured conductivity ratios fall below the theoretical curve, but this would be expected if effects other than mass difference contribute to the point-defect scattering.

Parrott (1963) and Abeles (1963) considered the same problem in more detail, taking into account N-processes by Callaway's method. They both reached the same conclusions and showed that their expressions became the same as that of Klemens when scattering by defects was much more important than scattering by U-processes. Abeles used the more refined expression to interpret the experimental results of Abeles *et al.* (1962), Steele and Rosi (1958), and Ioffe and Ioffe (1955). The good agreement between the calculated and observed variations of conductivity with composition for Ge–Si alloys has already been shown in Fig. 8.4.

13.2.2. Doped semiconductors

13.2.2(a). High temperatures. Beers *et al.* (1962) studied the effect of doping on the high-temperature thermal conductivity of germanium. For concentrations of arsenic and gallium up to $3 \times 10^{25}\,\text{m}^{-3}$ the thermal conductivity is reduced because the dopants and charge carriers scatter phonons and do not themselves carry enough heat to compensate for this reduction in phonon conductivity. The same behaviour has been observed in the III–V compound GaAs for both n- and p-type doping. On the other hand, for the III–V compounds InAs and InSb heavy doping with p-type impurities decreases the thermal conductivity, while doping with n-type impurities increases the conductivity. The mobility of the electrons is higher than that of the holes, so that the charge carriers make a contribution to the conductivity in n-type material which can exceed the decrease in phonon conductivity caused by the presence of the impurities.

Measurements on the IV–VI compounds PbSe and PbTe (Devyatkova and Smirnov 1960, Devyatkova, Petrov, and Smirnov 1961) and the V–VI compounds Bi_2Se_3 (Hashimoto 1958) and Bi_2Te_3 (Goldsmid 1958) at temperatures above 100 K show that the thermal conductivity increases with doping for both n- and p-type material.

176   SEMICONDUCTORS

In semiconductors for which the electrical conductivity is never negligible, it is possible to deduce what the lattice conductivity in pure material would be in the absence of scattering by charge carriers. At a particular temperature the variation of thermal conductivity with electrical conductivity as the doping is changed can be extrapolated to zero electrical conductivity. On the other hand, it is difficult to separate the measured conductivity of a given specimen into phonon and charge-carrier contributions, since the lattice conductivity is reduced both by the impurities and by the charge carriers to which they give rise.

13.2.2(b). Low temperatures. The effect of doping on the thermal conductivity of silicon and germanium at low temperatures has been studied by several authors. Carruthers, Geballe, Rosenberg, and Ziman (1957) found large reductions in the low-temperature conductivity of germanium and for certain dopings a very rapid variation of conductivity with temperature (κ varying even faster than T^{-4}), implying that, if this is a phonon conductivity, the mean free path *decreases* with decreasing temperature. Such behaviour was explained by Ziman (1956b, 1957) in terms of scattering of phonons by electrons in impurity bands.

If the relation between electron energy E and wave-number k is parabolic, with effective mass m_e^*, then in an N-process involving a phonon of wave vector $\mathbf{q}$ conservation of wave-vector and energy require

$$\mathbf{k}_1 + \mathbf{q} = \mathbf{k}_2$$

and

$$\hbar^2 k_1^2 / 2m_e^* + \hbar q v = \hbar^2 k_2^2 / 2m_e^*$$

assuming that the phonon velocity shows no dispersion. The minimum value of k_1 which satisfies both these equations for scattering of phonon q occurs when k_1 and q are collinear, and is given by $k_0 = |\frac{1}{2}q - m_e^* v/\hbar|$.

When $T \to 0$ and the dominant values of q are small enough for $\frac{1}{2}q$ to be negligible compared with $m_e^* v/\hbar$, k_0 is equal to $m_e^* v/\hbar$ and the corresponding minimum electron energy is $k_B T_s = \frac{1}{2}m_e^* v^2$. In an ordinary metal the electron distribution at normal and low temperatures is highly degenerate and only electrons within an energy range of $\sim k_B T$ on either side of the Fermi level can scatter phonons; $T_F > T_s$ and the lower limit of energy corresponding to k_0 is no restriction on the scattering. The phonon mean free path is inversely proportional to the number of effective scatterers and is thus proportional to $1/T$. If, however, the Fermi temperature is lower than T_s, the number of electrons with energies above the threshold decreases exponentially with decreasing temperature as $T \to 0$ and the phonon mean free path increases exponentially.

If $T \gg T_s$, the dominant phonons have wave-numbers much greater than $m_e^* v/\hbar$, and the threshold value of k_0 is nearly equal to $\frac{1}{2}q$. As the temperature increases, the electron distribution broadens with the maximum moving to higher values of k. As the threshold k_0 for scattering the dominant phonons with wave-vector q is proportional to q and thus to the temperature, the number of electrons with $k > k_0$ decreases and the phonon mean free path increases with increasing temperature. The limiting variation of the mean free path with temperature is as $T^{\frac{5}{2}}$.

This theory can account qualitatively for the rapid temperature variation found in the conductivity of one of the intermediately doped p-type specimens measured by Carruthers *et al.* Although at the temperatures concerned the number of electrons excited from the valence band into the impurity levels would be vanishingly small, the combination of a small effective mass for the carriers and a large dielectric constant for the medium can produce an overlapping of the impurity levels into a band. Conduction can take place within such a band and phonons can be scattered by electrons.

Carruthers *et al.* estimate that for the specimen with $2 \cdot 3 \times 10^{22}$ carriers/m^3, $T_F \sim 10 T_s$. The mean free path should decrease with decreasing temperature for $T \gtrsim 3 T_s$. T_s was estimated to be $\sim 0 \cdot 4$ K, so that l_e^p

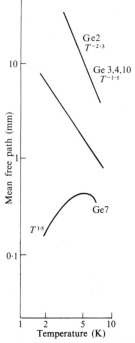

FIG. 13.3. Phonon mean free paths due to the additional scattering mechanism in specimens of germanium doped with indium. For the lowest curve the carrier concentration is $2 \cdot 3 \times 10^{22}$ carriers/m^3. The upper curves are for purer specimens. (After Carruthers *et al.* 1956.)

should decrease down to the lowest temperature of the measurements (2 K). From Fig. 13.3 it can be seen that the mean free path decreases below 5 K, corresponding to a conductivity varying faster than T^4 (above this temperature the mean free path varies in the normal way and the conductivity is proportional to a power of T less than 3).

In none of the specimens measured by Carruthers *et al.* was the electronic conductivity itself so large that it formed a measurable contribution to the overall conductivity, as is found at higher temperatures. For the highest doping (10^{25} carriers/m^3) the semiconductor behaved as a metal with rather few electrons, but T_F was greater than both T and T_s. The scattering of phonons additional to boundary scattering was found to be represented by a mean free path proportional to $T^{-1.4}$, rather than the theoretical variation of T^{-1}.

Measurements on doped GaAs, GaSb, and InSb at low temperatures show various temperature dependences of the thermal conductivity depending on the doping level, and such behaviour has been attributed to scattering of phonons by electrons in bound states (Holland 1964, Challis, Cheeke, and Williams 1965).

REFERENCES

General References

THE following books give more details on various aspects of the subject than was appropriate here.

Ziman, J. M. (1960). *Electrons and phonons.* Clarendon Press, Oxford.
   This covers the theory of transport phenomena in solids in great detail.
Parrott, J. E., and Stuckes, A. D. (1975). *Thermal conductivity of solids.* Pion, London.
   This is at a comparable level to the present work but with different emphases. More details are given about methods of measurement and about technological materials, and correspondingly less detail is given of the matter common to both books.
Tye, R. P. (ed.) (1969). *Thermal conductivity.* Academic Press, London.
   Two volumes containing 12 specialized chapters written by different authors. There is a chapter each on the theory of thermal conductivity in solids and in fluids, and there is much about experimental methods.
Drabble, J. R., and Goldsmid, H. J. (1961). *Thermal conduction in semiconductors.* Pergamon Press, Oxford.
   Specialized in the manner described by the title.

Abeles, B. (1963). *Phys. Rev.* **131,** 1906.
Abeles, B., Beers, D. S., Cody, G. D., and Dismukes, J. P. (1962). *Phys. Rev.* **125,** 44.
Ackerman, C. C., and Guyer, R. A. (1968). *Ann. Phys.* **50,** 128.
Ackerman, M. W. (1972). *Phys. Rev.* **B5,** 2751.
Akhieser, A. (1939). *Fiz. Zh.* **1,** 277.
Anderson, A. C., and Malinowski, M. E. (1972). *Phys. Rev.* **B5,** 3199.
Anderson, A. C., and Smith, S. C. (1973). *J. Physics Chem. Solids* **34,** 111.
Anderson, B. R., and Challis, L. J. (1975). *J. Phys. C solid st. Phys.* **8,** 1475; 1484; 1495.
Anderson, P. W., Halperin, B. I., and Varma, C. M. (1972). *Phil. Mag.* **25,** 1.
Anderson, V. C. (1950). *J. acoust. Soc. Am.* **22,** 426.
Ångström, A. J. (1861). *Annln. Phys.* **114,** 513.
Ashworth, T., Johnson, L. R., Hsiung, C. Y., and Kreitman, M. M. (1973). *Cryogenics* **13,** 34.
Bardeen, J. (1937). *Phys. Rev.* **52,** 688.
Bardeen, J., Cooper, L. N., and Schrieffer, J. R. (1957). *Phys. Rev.* **108,** 1175.
Bardeen, J., Rickayzen, G., and Tewordt, T. L. (1959). *Phys. Rev.* **113,** 982.
Baumann, F. C., and Pohl, R. O. (1967). *Phys. Rev.* **163,** 843.
Beck, H., Meier, P. F., and Thellung, A. (1974). *Phys. Status Solidi* (a) **24,** 11.
Beers, D. S., Cody, G. D., and Abeles, B. (1962). *Proc. Int. Conf. on the Physics of Semiconductors,* Exeter, p. 41. Institute of Physics and the Physical Society London.

REFERENCES

Benin, D. (1973). *Phys. Rev.* **A7**, 334.
Berman, R. (1949). *Phys. Rev.* **76**, 315.
Berman, R. (1953). *Adv. Phys.* **2**, 103.
Berman, R. (1961). In *Experimental cryophysics* (eds. F. E. Hoare, L. C. Jackson, and N. Kurti), chap. 10.10, p. 327. Butterworths, London.
Berman, R., Bounds, C. L., Day, C. R., *and* Sample, H. H. (1968). *Phys. Lett.* **26A**, 185.
Berman, R., Bounds, C. L., *and* Rogers, S. J. (1965). *Proc. R. Soc.* **A289**, 66.
Berman, R., *and* Brock, J. C. F. (1965). *Proc. R. Soc.* **A289**, 46.
Berman, R., Brock, J. C. F., *and* Huntley, D. J. (1963). *Phys. Lett.* **3**, 310.
Berman, R., Day, C. R., Goulder, D. P., *and* Vos, J. E. (1973). *J. Phys. C solid st. Phys.* **6**, 2119.
Berman, R., Foster, E. L., *and* Ziman, J. M. (1955). *Proc. R. Soc.* **A231**, 130.
Berman, R., Foster, E. L., *and* Ziman, J. M. (1956). *Proc. R. Soc.* **A237**, 344.
Berman, R., Nettley, P. T., Sheard, F. W., Spencer, A. N., Stevenson, R. W. H., *and* Ziman, J. M. (1959). *Proc. R. Soc.* **A253**, 403.
Berman, R., Simon, F. E., *and* Ziman, J. M. (1953). *Proc. R. Soc.* **A220**, 171.
Bilir, N., *and* Phillips, W. A. (1975). *Phil. Mag.* **32**, 113.
Black, M. A. (1973). *Am. J. Phys.* **41**, 691.
Blackman, M. (1935). *Phil. Mag.* **19**, 989.
Blatt, F. J. (1968). *Physics of electronic conduction in solids*, p. 303. McGraw-Hill, New York.
Bloch, F. (1930). *Z. Phys.* **59**, 208.
Böttger, H. (1974). *Phys. Status Solidi* (b), **62**, 9.
Brock, J. C. F., *and* Huntley, D. J. (1968). *Can. J. Phys.* **46**, 2231.
Bross, H., Seeger, A., *and* Haberkorn, R. (1963). *Phys. Status Solidi* **3**, 1126.
Callaway, J. (1959). *Phys. Rev.* **113**, 1046.
Carruthers, P. (1961). *Rev. Mod. Phys.* **33**, 92.
Carruthers, J. A., Geballe, T. H., Rosenberg, H. M., *and* Ziman, J. M. (1957). *Proc. R. Soc.* **A238**, 502.
Carwile, L. C. K., *and* Hoge, H. J. (1966). Tech. Rep. No. 67-7-PR, U.S. Army, Natick Labs.
Casimir, H. B. G. (1938). *Physica* **5**, 495.
Challis, L. J., Cheeke, J. D., *and* Williams, D. J. (1965). *Proc. 9th Int. Conf. on Low Temperature Physics*, p. 1145. Plenum Press, New York.
Childs, G. E., Ericks, L. J., *and* Powell, R. L. (1973). *Thermal conductivity of solids at room temperature and below.* National Bureau of Standards Monograph 131, U.S. Department of Commerce, Washington, D.C.
Choy, C. L., *and* Greig, D. (1975). *J. Phys. C solid st. Phys.* **8**, 3121.
Choy, C. L., Hunt, R. G., *and* Salinger, G. L. (1970). *J. Chem. Phys.* **52**, 3629.
Choy, C. L., Salinger, G. L., *and* Chiang, Y. C. (1970). *J. appl. Phys.* **41**, 597.
Cieloszyk, G. S., Cruz, M. T., *and* Salinger, G. L. (1973). *Cryogenics* **13**, 718.
Clayton, F., *and* Batchelder, D. N. (1973). *J. Phys. C solid st. Phys.* **6**, 1213.
Cook, J. G., van der Meer, M. P., *and* Laubitz, M. J. (1972). *Can. J. Phys.* **50**, 1386.
Danielson, G. C., *and* Sidles, P. H. (1969). In *Thermal conductivity*, (ed. R. P. Tye), vol. 2, chap. 3, p. 149. Academic Press, London.
Davydov, B., and Shmushkevitch, I. (1940). *Usp. fiz. Nauk SSSR* **24**, 21.
Day, C. R. (1970). Thesis, Oxford University.
Debye, P. (1914). *Vorträge über die kinetische Theorie der Materie und der Elektrizität*, p. 17. Teubner, Leipzig and Berlin.

Devyatkova, E. D., Petrov, A. V., *and* Smirnov, I. A. (1961). *Fizika tverd Tela.* **3**, 1338; *Sov. Phys. Solid St.* **3**, 970 (1961).
Devyatkova, E. D., *and* Smirnov, I. A. (1960). *Fizika tverd. Tela* **2**, 1984; *Sov. Phys. Solid St.* **2**, 1786 (1960).
Devyatkova, E. D., *and* Stil'bans, L. S. (1952). *Zh. tekh. Fiz. USSR* **22**, 968.
Douglass, R. L. (1963). *Phys. Rev.* **129**, 1132.
Drabble, J. R., *and* Goldsmid, H. J. (1961). *Thermal conduction in semiconductors*, p. 115. Pergamon Press, Oxford.
Dreyfus, B., Fernandes, N. C., *and* Maynard, R. (1968). *Phys. Lett.* **26A**, 647.
Dreyfus, B., Lacaze, A., *and* Zadworny, F. (1962). *C.R. Acad. Sci. Paris* **254**, 3337.
Drude, P. (1900). *Annln. Phys.* **1**, 566.
Drude, P. (1902). *Annln. Phys.* **7**, 687.
Dugdale, J. S., *and* MacDonald, D. K. C. (1955). *Phys. Rev.* **98**, 1751.
Elliott, R. J., *and* Parkinson, J. B. (1967). *Proc. phys. Soc.* **92**, 1024.
Eucken, A., *and* Kuhn, G. (1928). *Z. Phys. Chem.* **134**, 193.
Flubacher, P., Leadbetter, A. J., Morrison, J. A., *and* Stoicheff, B. P. (1959). *J. Phys. Chem. Solids* **12**, 53.
Flynn, D. R. (1969). In *Thermal conductivity* (ed. R. P. Tye), vol. 1, chap. 5, p. 241. Academic Press, London.
Fox, J. N. (1971). Ph.D. Thesis, Wesleyan University, Middletown, Connecticut, U.S.A.
Friedberg, S. A., *and* Douthett, D. (1958). *Physica* **24**, S176.
Friedberg, S. A., *and* Harris, E. D. (1963). *Proc. 8th Int. Conf. on Low Temperature Physics*, London, 1962 (ed. R. O. Davies), p. 302. Butterworths, London.
Fulkerson, W., Moore, J. P., Williams, R. K., Graves, R. S., *and* McElroy, D. L. (1968). *Phys. Rev.* **167**, 765.
Garber, J. A., *and* Granato, A. V. (1970). *J. Phys. Chem. Solids* **31**, 1863.
Garber, M., Scott, B. W., *and* Blatt, F. J. (1963). *Phys. Rev.* **130**, 2188.
Garland, J. C., *and* Bowers, R. (1968). *Phys. Rev. Lett.* **21**, 1007.
Garrett, K. W., *and* Rosenberg, H. M. (1972). *Proc. 4th Int. Cryogenic Engineering Conf.*, Eindhoven, p. 267. IPC Science and Technology Press, Guildford, Surrey.
Geballe, T. H. *and* Hull, G. W. (1955). *Conf. de Physique des Basses Températures*, Paris, p. 460. Annexe 1955-3 Supplément au Bulletin de l'Institut International du Froid, 177 Boulevard Malesherbe, Paris 17e.
Geballe, T. H., *and* Hull, G. W. (1958). *Phys. Rev.* **110**, 773.
Genzel, L. (1953). *Z. Phys.* **135**, 177.
Giles, M., *and* Terry, C. (1969). *Phys. Rev. Lett.* **22**, 882.
Glassbrenner, C. J., *and* Slack, G. A. (1964). *Phys. Rev.* **134**, A1058.
Golding, B., Graebner, J. E., Halperin, B. I., *and* Schutz, R. J. (1973). *Phys. Rev. Lett.* **30**, 223.
Goldsmid, H. J. (1958). *Proc. phys. Soc.* **72**, 17.
Goulder, D. P. (1973). Thesis, Oxford University.
Granato, A. (1958). *Phys. Rev.* **111**, 740.
Green, A. *and* Cowles, L. E. J. (1960). *J. Sci. Instrum.* **37**, 349.
Gurzhi, R. N. (1964). *Zh. Eksper. teor. Fiz.* **46**, 719; *Sov. Phys. JETP* **19**, 490 (1964).
Guyer, R. A., *and* Krumhansl, J. A. (1966). *Phys. Rev.* **148**, 766, 778.
de Haas, W. J., *and* Biermasz, Th. (1935). *Physica* **2**, 673.
Hamilton, R. A. H., *and* Parrott, J. E. (1969). *Phys. Rev.* **178**, 1284.

Harley, R. T., McClintock, P. V. E., *and* Rosenberg, H. M. (1969). *Phys. Lett.* **28A,** 469.
Hashimoto, K. (1958). *Mem. Fac. Sci. Kyusyu Univ.* **B2,** 187.
Heer, C. V., *and* Daunt, J. G. (1949). *Phys. Rev.* **76,** 854.
Herring, C. (1954). *Phys. Rev.* **95,** 954.
Herring, C. (1967). *Phys. Rev. Lett.* **19,** 167 (errata p. 684).
Hogan, E. M., Guyer, R. A., *and* Fairbank, H. A. (1969). *Phys. Rev.* **185,** 356.
Holland, M. G. (1964). *Phys. Rev.* **134,** A471.
Howling, D. H., Mendoza, E., *and* Zimmerman, J. G. (1955). *Proc. R. Soc.* **A229,** 86.
Hunklinger, S., Arnold, W., *and* Stein, S. (1973). *Phys. Lett.* **45A,** 311.
Hunklinger, S., Arnold, W., Stein, S., Nava, R., *and* Dransfeld, K. (1972). *Phys. Lett.* **42A,** 253.
Hunklinger, S., Piché, L., Lasjaunias, J. C., *and* Dransfeld, K. (1975). *J. Phys. C. Solid St. Phys.* **8,** L423.
Hust, J. G., *and* Sparks, L. L. (1973). *Natn. Bur. Stand. Tech. Note* 634.
Ioffe, A. F., Airapetyants, S. V., Ioffe, A. V., Kolomoets, N. V., *and* Stil'bans, L. S. (1956). *Dokl. Akad. Nauk SSSR* **106,** 981; *Sov. Phys. Doklady* **1,** 132.
Ioffe, A. V., *and* Ioffe, A. F. (1955). *Izvest. Akad. Nauk SSSR* **20,** 76; *Bull. Acad. Sci. USSR Phys. Ser.* **20,** 66 (1956).
Jackson, H. E., *and* Walker, C. T. (1971). *Phys. Rev.* **B3,** 1428.
Jacobsen, E. H., *and* Stevens, K. W. H. (1963). *Phys. Rev.* **129,** 2036.
Jones, H. D. (1970). *Phys. Rev.* **A1,** 71.
Julian, C. L. (1965). *Phys. Rev.* **137,** A128.
Kapoor, A., Rowlands, J. A., *and* Woods, S. B. (1974). *Phys. Rev.* **B9,** 1223.
Karamargin, M. C., Reynolds, C. A., Lipschultz, F. P., *and* Klemens, P. G. (1972a). *Phys. Rev.* **B5,** 2856.
Karamargin, M. C., Reynolds, C. A., Lipschultz, F. P., *and* Klemens, P. G. (1972b). *Phys. Rev.* **B6,** 3624.
Kemp, W. R. G., Klemens, P. G., Sreedhar, A. K., *and* White, G. K. (1956). *Proc. R. Soc.* **A233,** 480.
Kemp, W. R. G., Klemens, P. G., *and* Tainsh, R. J. (1957). *Aust. J. Phys.* **10,** 454.
Kemp, W. R. G., Klemens, P. G., *and* Tainsh, R. J. (1959). *Phil. Mag.* **4,** 845.
Kimber, R. M., *and* Rogers, S. J. (1973). *J. Phys. C solid st. Phys.* **6,** 2279.
Kittel, C. (1949). *Phys. Rev.* **75,** 972.
Kittel, C. (1971). *Introduction to solid state physics* (4th edn.). John Wiley, New York.
Klein, A. H., Shanks, H. R., *and* Danielson, G. C. (1963). *Proc. 3rd Conf. on Thermal Conductivity*, Gatlinburg, Tennessee (unpublished).
Klemens, P. G. (1951). *Proc R. Soc.* **A208,** 108.
Klemens, P. G. (1954). *Aust. J. Phys.* **7,** 64.
Klemens, P. G. (1955). *Proc. phys. Soc.* **A68,** 1113.
Klemens, P. G. (1958). *Solid State Physics* (eds. F. Seitz and D. Turnbull), vol. 7, p. 1. Academic Press, New York.
Klemens, P. G. (1960). *Phys. Rev.* **119,** 507.
Klemens, P. G. (1965). In *Physics of non-crystalline solids* (ed. J. A. Prins), p. 162. North-Holland, Amsterdam.
Klemens, P. G. (1969). In *Thermal conductivity* (ed. R. P. Tye), vol. 1, chap. 1, p. 1. Academic Press, London.
Klemens, P. G., *and* Ecsedy, D. J. (1976). *Proc. 2nd Int. Conf. on Phonon*

REFERENCES

Scattering in Solids, Nottingham, 1975 (eds. L. J. Challis, V. W. Rampton, and A. F. G. Wyatt), p. 367. Plenum Press, London.
de Klerk, J., *and* Klemens, P. G. (1966). *Phys. Rev.* **147,** 585.
Knudsen, M. (1909). *Annln. Phys.* **28,** 75.
Kopp, J., *and* Slack, G. A. (1971). *Cryogenics* **11,** 22.
Kopylov, V. N., *and* Mezhov-Deglin, L. P. (1971). *Zh. eksper. teor. Fiz. Pis. Red.* **14,** 32; *Sov. Phys. JETP Lett.* **14,** 21 (1971).
Kopylov, V. N., *and* Mezhov-Deglin, L. P. (1973). *Zh. eksper. teor. Fiz.* **65,** 720; *Sov. Phys. JETP* **38,** 357 (1974).
Krumhansl, J. A. (1964). *Proc. Int. Conf. on Lattice Dynamics*, Copenhagen, 1963, p. 523. Pergamon Press, Oxford.
Krumhansl, J. A., *and* Matthew, J. A. D. (1965). *Phys. Rev.* **140,** A1812.
Krupskii, I. N., *and* Manzhely, V. G. (1967). *Phys. Status Solidi,* **24,** K53.
Krupskii, I. N., *and* Manzhelii, V. G. (1968). *Zh. eksper. teor. Fiz.* **55,** 2075; *Sov. Phys. JETP* **28,** 1097 (1969).
Laubitz, M. J. (1969). In *Thermal conductivity* (ed. R. P. Tye), vol. 1, chap 3, p. 111. Academic Press, London.
Laubitz, M. J., *and* Cook, J. G. (1972). *Phys. Rev.* **B6,** 2082.
Lawson, A. W. (1957). *J. Physics Chem. Solids* **3,** 155.
Lawson, D. T., *and* Fairbank, H. A. (1973). *J. low temp. Phys.* **11,** 363.
Leadbetter, A. J. (1968). *Physics Chem. Glasses* **9,** 1.
Leadbetter, A. J. (1969). *J. chem. Phys.* **51,** 779.
Leadbetter, A. J. (1972). *Proc. Int. Conf. on Phonon Scattering in Solids*, Paris (ed. H. J. Albany), p. 338, Service de Documentation du CEN Saclay.
Leaver, A. D. W., *and* Charsley, P. (1971). *J. Phys. F Metal. Phys.* **1,** 28.
Leibfried, G., *and* Schlömann, E. (1954). *Nachr. Akad. Wiss. Göttingen* II **a(4),** 71.
Lindenfeld, P., *and* Pennebaker, W. B. (1962). *Phys. Rev.* **127,** 1881.
Little, W. A. (1959). *Can. J. Phys.* **37,** 334.
Logachev, Yu. A., *and* Vasil'ev, L. N. (1973). *Fizika tverd. Tela* **15,** 1612; *Sov. Phys. Solid State* **15,** 1081.
Lomer, J. N., *and* Rosenberg, H. M. (1959). *Phil. Mag.* **4,** 467.
Lorenz, L. (1881). *Annln. Phys.* **13,** 422.
Love, W. F. (1973). *Phys. Rev. Lett.* **31,** 822.
Lüthi, B. (1962). *J. Physics Chem. Solids* **23,** 35.
McClintock, P. V. E., Morton, I. P., Orbach, R., *and* Rosenberg, H. M. (1967). *Proc. R. Soc.* **A298,** 359.
McCurdy, A. K., Maris, H. J., *and* Elbaum, C. (1970). *Phys. Rev.* **B2,** 4077.
Macdonald, D. K. C., White, G. K., *and* Woods, S. B. (1956). *Proc. R. Soc.* **A235,** 358.
McElroy, D. L., *and* Moore, J. P. (1969). In *Thermal conductivity* (ed. R. P. Tye), vol. 1, chap. 4, p. 185. Academic Press, London.
Makinson, R. E. B. (1938). *Proc. Camb. phil. Soc.* **34,** 474.
Martin, J. J., Shanks, H. R., *and* Danielson, G. C. (1968). *Proc. 7th Thermal Conductivity Conf.* 1967, (eds. D. R. Flynn and B. A. Penry), p. 381. National Bureau of Standards Special Publications No. 302.
Matthiessen, A. (1862). *Rep. Br. Ass.* **32,** 144.
Meaden, G. T. (1966). *Electrical resistance of metals*. Heywood Books, London.
Men', A. A., *and* Sergeev, O. A. (1973). *High Temp.–High Press.* **5,** 19.
Mendelssohn, K. *and* Schiffman, C. A. (1960). *Proc. R. Soc.* **A255,** 199.
Metcalfe, M. J., *and* Rosenberg, H. M. (1972). *J. Phys. C solid st. Phys.* **5,** 459.

Mezhov-Deglin, L. P. (1964). *Zh. eksper. teor. Fiz.* **46,** 1926; *Sov. Phys. JETP* **19,** 1297 (1964).
Mezhov-Deglin, L. P. (1965). *Zh. eksper. teor. Fiz.* **49,** 66; *Sov. Phys. JETP* **22,** 47 (1966).
Mezhov-Deglin, L. P. (1967). *Zh. eksper. teor. Fiz.* **52,** 866; *Sov. Phys. JETP* **25,** 568 (1967).
Mitchell, M. A., Klemens, P. G., *and* Reynolds, C. A. (1971). *Phys. Rev.* **B3,** 1119.
Moore, J. P., McElroy, D. L., *and* Barisoni, M. (1966). *Proc. 6th Thermal Conductivity Conference*, p. 737. Air Force Materials Lab., Dayton, Ohio.
Moore, J. P., Williams, R. K., *and* Graves, R. S. (1975). *Phys. Rev.* **B11,** 3107.
Morgan, G. J., *and* Smith, D. (1974). *J. Phys. C solid st. Phys.* **7,** 649.
Morton, I. P., *and* Rosenberg, H. M. (1962). *Phys. Rev. Lett.* **8,** 200.
Moss, M. (1965). *J. Appl. Phys.* **36,** 3308.
Moss, M. (1966). *J. Appl. Phys.* **37,** 4168.
Mueller, K. H., *and* Fairbank, H. A. (1971). *Proc. 12th Int. Conf. on Low Temperature Physics*, p. 135. Academic Press of Japan, Tokyo.
Nabarro, F. R. N. (1951). *Proc. R. Soc.* **A209,** 278.
Narayanamurti, V., Seward, W. D., *and* Pohl, R. O. (1966). *Phys. Rev.* **148,** 481.
Ninomiya, T. (1968). *J, phys. Soc. Japan* **25,** 830.
O'Hara, S. G., *and* Anderson, A. C. (1974a). *Phys. Rev.* **B9,** 3730.
O'Hara, S. G., *and* Anderson, A. C. (1974b). *Phys. Rev.* **B10,** 574.
Ohashi, K. (1968). *J. Phys. Soc. Japan* **24,** 437.
Olsen, J. L. (1952). *Proc. phys. Soc.* **A65,** 518.
Parrott, J. E. (1963). *Proc. phys. Soc.* **81,** 726.
Peierls, R. (1929). *Annln. Phys.* **3,** 1055.
Peierls, R. E. (1955). *Quantum theory of solids*, p. 45. Clarendon Press, Oxford.
Phillips, W. A. (1972). *J. low temp. Phys.* **7,** 351.
Piché, L., Maynard, R., Hunklinger, S., *and* Jäckle, J. (1974). *Phys. Rev. Lett.* **32,** 1426.
Pippard, A. B. (1955). *Phil. Mag.* **46,** 1104.
Pohl, R. O. (1962). *Phys. Rev. Lett.* **8,** 481.
Pohl, R. O., Love, W. F., *and* Stephens, R. B. (1974). *Proc. 5th Int. Conf. on Amorphous and Liquid Semiconductors*, Garmisch-Partenkirchen, 1973, (eds. J. Stuke and W. Brenig) p. 1121. Taylor and Francis, London.
Pomeranchuk, I. (1941). *J. Phys. USSR* **4,** 259.
Powell, R. W. (1957). *J. sci. Instrum.* **34,** 485.
Powell, R. W. (1969). *Contemp. Phys.* **10,** 579.
Price, P. J. (1955). *Phil. Mag.* **46,** 1252.
Ranninger, J. (1965). *Phys. Rev.* **140,** A2031.
Rayleigh, Lord (1896). *Theory of sound* (2nd edn.), vol. (ii). Dover Publications, New York (1945).
Reese, W., *and* Steyert, W. A. (1962). *Rev. sci. Instrum.* **33,** 43.
Reese, W., *and* Tucker, J. E. (1965). *J. chem. Phys.* **43,** 105.
Rosenberg, H. M., *and* Sujak, B. (1960). *Phil. Mag.* **5,** 1299.
Roufosse, M., *and* Klemens, P. G. (1973). *Phys. Rev.* **B7,** 5379.
Saint-James, D., Sarma, G., *and* Thomas, E. J. (1969). *Type II superconductivity*, Pergamon Press, Oxford.
Sato, H. (1955). *Prog. theor. Phys. (Kyoto)* **13,** 119.
Savvides, N., *and* Goldsmid, H. J. (1972). *Phys. Lett.* **41A,** 193.
Savvides, N., *and* Goldsmid, H. J. (1973). *J. Phys. C solid st. Phys.* **6,** 1701:
Savvides, N., *and* Goldsmid, H. J. (1974). *Phys. Status Solidi* (b) **63,** K89.

REFERENCES

Schoek, G. (1962). *J. appl. Phys.* **33,** 1745.
Schwartz, J. W., and Walker, C. T. (1966). *Phys. Rev. Lett.* **16,** 97.
Schwartz, J. W., and Walker, C. T. (1967a). *Phys. Rev.* **155,** 959.
Schwartz, J. W., and Walker, C. T. (1967b). *Phys. Rev.* **155,** 969.
Seward, W. D., and Narayanamurti, V. (1966). *Phys. Rev.* **148,** 463.
Simons, S. (1960). *Proc. phys. Soc.* **76,** 458.
Slack, G. A. (1957). *Phys. Rev.* **105,** 832.
Slack, G. A. (1972a). *Proc. Int. Conf. on Phonon Scattering in Solids*, (ed. H. J. Albany), p. 24. Service de Documentation du CEN Saclay.
Slack, G. A. (1972b). *Phys. Rev.* **B6,** 3791.
Slack, G. A. (1973). *J. Physics Chem. Solids* **34,** 321.
Slack, G. A. (1977). To be published in *Solid state physics* (eds. H. Ehrenreich, F. Seitz, and D. Turnbull). Academic Press, New York.
Slack, G. A., and Glassbrenner, C. J. (1960). *Phys. Rev.* **120,** 782.
Slack, G. A., and Newman, R. (1958). *Phys. Rev. Lett.* **1,** 359.
von Smoluchowski, M. (1910). *Annln. Phys.* **33,** 1559.
Sondheimer, E. H. (1950). *Proc. R. Soc.* **A203,** 75.
Sproull, R. L., Moss, M., and Weinstock, H. (1959). *J. appl. Phys.* **30,** 334.
Srivastava, G. P., and Verma, G. S. (1973). *Can. J. Phys.* **51,** 223.
Steele, M. C., and Rosi, F. D. (1958). *J. appl. Phys.* **29,** 1517.
Steigmeier, E. F. (1969). In *Thermal conductivity* (ed. R. P. Tye) vol. 2, chap. 4, p. 203. Academic Press, London.
Stephens, R. B. (1973). *Phys. Rev.* **B8,** 2896.
Stephens, R. B. (1976). *Phys. Rev.* **B13,** 852.
Stuckes, A. D. (1960). *Phil. Mag.* **5,** 84.
Sussmann, J. A., and Thellung, A. (1963). *Proc. phys. Soc.* **81,** 1122.
Suzuki, T., and Suzuki, H. (1972). *J. phys. Soc. Japan* **32,** 164.
Tainsh, R. J., and White, G. K. (1962). *J. Physics Chem. Solids* **23,** 1329.
Taylor, A., Albers, H. R., and Pohl, R. O. (1965). *J. appl. Phys.* **36,** 2270.
Thacher, P. D. (1965). Ph.D. Thesis, Cornell University. (See *Phys. Rev.* **156,** 975 (1967).)
Touloukian, Y. S. (ed.) (1970). *Thermophysical properties of matter. The TPRC data series*, vols. 1 and 2. IFI/Plenum Press, New York.
Turk, L. A., and Klemens, P. G. (1974). *Phys. Rev.* **B9,** 4422.
Wagner, C. M. (1963). *Phys. Rev.* **131,** 1443.
Waldorf, D. L., Boughton, R. I., Yaqub, M., and Zych, D. (1972). *J. low temp. Phys.* **9,** 435.
Walker, C. T. (1963). *Phys. Rev.* **132,** 1963.
Walker, C. T., and Pohl, R. O. (1963). *Phys. Rev.* **131,** 1433.
Walton, D. (1967). *Phys. Rev.* **157,** 720.
Walton, D., and Lee, E. J. (1967). *Phys. Rev.* **157,** 724.
Walton, D., Rives, J. E., and Khalid, Q. (1973). *Phys. Rev.* **B8,** 1210.
White, G. K. (1969). In *Thermal conductivity* (ed. R. P. Tye), vol. 1, chap. 2, p. 69. Academic Press, London.
White, G. K. (1975). *Phys. Rev. Lett.* **34,** 204.
White, G. K., and Tainsh, R. J. (1967). *Phys. Rev. Lett.* **19,** 165.
White, G. K., and Woods, S. B. (1959). *Phil. Trans. R. Soc.* **A251,** 273.
Whitworth, R. W. (1958). *Proc. R. Soc.* **A246,** 390.
Wiedemann, G., and Franz, R. (1853). *Annln. Phys.* **89,** 497.
Wilson, A. H. (1937). *Proc. Camb. phil. Soc.* **33,** 371.
Wilson, A. H. (1953). *The theory of metals* (2nd edn.). Cambridge University Press, London.

Worlock, J. M. (1966). *Phys. Rev.* **147,** 636.
Ying, C. F., *and* Truell, R. (1956). *J. appl. Phys.* **27,** 1086.
Yussouff, M., *and* Mahanty, J. (1966). *Proc. phys. Soc.* **87,** 689.
Yussouff, M., *and* Mahanty, J. (1967). *Proc. phys. Soc.* **90,** 519.
Zaitlin, M. P., *and* Anderson, A. C. (1974*a*). *Phys. Rev.* **B10,** 580.
Zaitlin, M. P., *and* Anderson, A. C. (1974*b*). *Phys. Rev. Lett.* **33,** 1158.
Zaitlin, M. P., *and* Anderson, A. C. (1975). *Phys. Rev.* **B12,** 4475.
Zavaritskii, N. V. (1957). *Zh. eksper. teor. Fiz.* **33,** 1085; *Sov. Phys. JETP* **6,** 837 (1958).
Zavaristskii, N. V. (1958). *Zh. eksper. teor. Fiz.* **34,** 1116; *Sov. Phys. JETP* **7,** 773 (1958).
Zeller, R. C., *and* Pohl, R. O. (1971). *Phys. Rev.* **B4,** 2029.
Ziman, J. M. (1954). *Proc. R. Soc.* **A226,** 436.
Ziman, J. M. (1956*a*). *Can. J. Phys.* **34,** 1256.
Ziman, J. M. (1956*b*). *Phil. Mag.* **1,** 191; corrected *Phil. Mag.* **2,** 292 (1957).
Ziman, J. M. (1960). *Electrons and phonons.* Clarendon Press, Oxford.
Ziman, J. M. (1963). *Electrons in metals,* Taylor & Francis, London. (Originally a series of articles in *Contemp. Phys.* (1962).)
Zimmerman, J. E. (1959). *J. Physics Chem. Solids* **11,** 299.

AUTHOR INDEX

Abeles, B., 91, 170, 175
Ackerman, C. C., 89
Ackerman M. W., 163
Airapetyants, S. V., 174
Akhieser, A., 47
Albers, H. R., 101
Anderson, A. C., 66, 79, 101, 102, 112, 113, 114, 164
Anderson, B. R., 98
Anderson, P. W., 111
Anderson, V. C., 77, 100, 101
Ångström, A. J., 7
Arnold, W., 112
Ashworth, T., 107

Bardeen, J., 137, 165
Barisoni, M., 158
Batchelder, D. N., 48–9
Baumann, F. C., 76
Beck, H., 1, 21
Beers, D. S., 91, 170, 175
Benin, D., 60
Berman, R., 5, 40, 58, 60, 65, 66, 67, 83–6, 87–8, 89, 95, 97, 98, 109
Biermasz, Th., 61, 62
Bilir, N., 111
Black, M. A., 47
Blackman, M., 53, 173
Blatt, F. J., 115, 155, 161, 171
Bloch, F., 133
Böttger, H., 104
Boughton, R. L., 151
Bounds, C. L., 87–8
Bowers, R., 150
Brock, J. C. F., 83–6, 95, 97, 98
Bross, H., 78

Callaway, J., 28, 36–40, 43, 44, 81, 82, 83, 91
Carruthers, J. A., 176–8
Carruthers, P., 73, 78, 101, 103
Carwile, L. C. K., 105, 107
Casimir, H. B. G., 61–62, 64, 66
Challis, L. J., 98, 178
Charsley, P., 163
Cheeke, J. D., 178
Chiang, Y. C., 107
Childs, G. E., 143

Choy, C. L., 107–8, 110
Cieloszyk, G. S., 107
Clayton, F., 48–9
Cody, G. D., 91, 170, 175
Cook, J. G., 132, 146–7
Cooper, L. N., 165
Cowles, L. E. J., 8
Cruz, M. T., 107

Danielson, G. C., 9, 173
Daunt, J. G., 168
Davydov, B., 172
Day, C. R., 44, 60, 87–8
Debye, P., 13, 29
Devyatkova, E. D., 81, 175
Dismukes, J. P., 91, 170, 175
Douglass, R. L., 99
Douthett, D., 98
Drabble, J. R., 172, 174, 175
Dransfeld, K., 112, 114
Dreyfus, B., 94, 110
Drude, P., 11, 116
Dugdale, J. S., 47

Ecsedy, D. J., 50, 170
Elbaum, C., 67
Elliott, R. J., 98
Ericks, L. J., 143
Eucken, A., 54–5

Fairbank, H. A., 60–1, 70, 71, 87–8, 89
Fernandes, N. C., 110
Flubacher, P., 110
Flynn, D. R., 7, 10
Foster, E. L., 58, 66, 67
Fox, J. N., 89
Franz, R., 11, 116
Friedberg, S. A., 98, 99
Fulkerson, W., 170

Garber, J. A., 79, 103, 164
Garber, M., 155, 161
Garland, J. C., 150
Garrett, K. W., 108
Geballe, T. H., 68, 83, 90, 176–8
Genzel, L., 106
Giles, M., 107
Glassbrenner, C. J., 90, 170–3
Golding, B., 112

Goldsmid, H. J., 68–9, 172, 174, 175
Goulder, D. P., 60, 89
Graebner, J. E., 112
Granato, A. V., 79, 101, 103, 164
Graves, R. S., 54, 170
Green, A., 8
Greig, D., 107–8
Gurzhi, R. N., 70
Guyer, R. A., 43–4, 60–1, 70, 71, 89

de Haas, W. J., 61, 62
Haberkorn, R., 78
Halperin, B. I., 111, 112
Hamilton, R. A. H., 66, 89–90
Harley, R. T., 97
Harris, E. D., 99
Hashimoto, K., 175
Heer, C. V., 168
Herring, C., 66, 68–9, 151
Hogan, E. M., 60–1, 70, 71, 89
Hoge, H. J., 105, 107
Holland, M. G., 173, 178
Howling, D. H., 8
Hsiung, C. Y., 107
Hull, G. W., 68, 83, 90
Hunklinger, S., 112, 114
Hunt, R. G., 110
Huntley, D. J., 95, 97, 98
Hust, J. G., 143, 152, 153, 158

Ioffe, A. F., 174, 175
Ioffe, A. V., 174, 175

Jäckle, J., 112
Jackson, H. E., 58, 59
Jacobsen, E. H., 98
Johnson, L. R., 107
Jones, H. D., 87
Julian, C. L., 47, 49, 50, 57

Kapoor, A., 163
Karamargin, M. C., 154, 157
Kemp, W. R. G., 153, 156, 157, 162
Khalid, Q., 99
Kimber, R. M., 86
Kittel, C., 19, 98, 104, 109, 138
Klein, A. H., 173
Klemens, P. G., 1, 41–2, 47, 50, 51, 57, 58, 70, 73, 75–6, 77, 78, 85, 90, 101, 103, 109, 130, 146, 148, 149, 153, 154, 156, 157, 162, 163, 170, 174–5
de Klerk, J., 85
Knudsen, M., 63
Kolomoets, N. V., 174
Kopp, J., 5
Kopylov, V. N., 72
Kreitman, M. M., 107

Krumhansl, J. A., 43–4, 70, 76, 94
Krupskii, I. N., 49
Kuhn, G., 54–5

Lacaze, A., 94
Lasjaunias, J. C., 114
Laubitz, M. J., 5, 132, 146–7
Lawson, A. W., 47
Lawson, D. T., 70, 87–8, 89
Leadbetter, A. J., 110, 111
Leaver, A. D. W., 163
Lee, E. J., 100
Leibfried, G., 23, 47, 57, 58
Lindenfeld, P., 139–40, 155, 159–61
Lipschultz, F. P., 154, 157
Little, W. A., 108
Logachev, Yu. A., 170
Lomer, J. N., 162
Lorenz, L., 12, 116
Love, W. F., 112, 113
Lüthi, B., 99

MacDonald, D. K. C., 47, 148–9
McClintock, P. V. E., 97, 98
McCurdy, A. K., 67
McElroy, D. L., 6, 9–10, 158, 170
Mahanty, J., 76
Makinson, R. E. B., 133
Malinowski, M. E., 79, 102
Manzhelii, V. G., 49
Maris, H. J., 67
Martin, J. J., 173
Matthew, J. A. D., 76
Matthiessen, A., 128
Maynard, R., 110, 112
Meaden, G. T., 148
van der Meer, M. P., 146–7
Meier, P. F., 1, 21
Men', A. A., 107
Mendelssohn, K., 168
Mendoza, E., 8
Metcalfe, M. J., 99
Mezhov-Deglin, L. P., 60, 70–1, 72
Mitchell, M. A., 163
Moore, J. P., 6, 9–10, 54, 158, 170
Morgan, G. J., 112
Morrison, J. A., 110
Morton, I. R., 94, 97, 98
Moss, M., 101
Mueller, K. H., 89

Nabarro, F. R. N., 78, 79
Narayanamurti, V., 92–3
Nava, R., 112
Nettley, P. T., 40, 89
Newman, R., 94
Ninomiya, T., 79, 103

AUTHOR INDEX

O'Hara, S. G., 164
Ohashi, K., 78, 103
Olsen, J. L., 167
Orbach, R., 97, 98

Parkinson, J. B., 98
Parrott, J. E., 66, 89–90, 175
Peierls, R., 20, 30, 57, 61
Pennebaker, W. B., 139–140, 155, 159–61
Petrov, A. V., 175
Phillips, W. A., 111, 112
Piché, L., 112, 114
Pippard, A. B., 139, 158
Pohl, R. O., 76, 81, 82, 90, 92, 93, 99, 101, 105, 106, 110, 111, 112, 113
Pomeranchuk, I., 50, 170
Powell, R. L., 143
Powell, R. W., 9, 146
Price, P. J., 172

Ranninger, J., 50, 170
Rayleigh, Lord, 74–5
Reese, W., 107, 168
Reynolds, C. A., 154, 157, 163
Rickayzen, G., 165
Rives, J. E., 99
Rogers, S. J., 86, 87–8
Rosenberg, H. M., 94, 97, 98, 99, 108, 162, 176–8
Rosi, F. D., 175
Roufosse, M., 47, 51
Rowlands, J. A., 163

Saint-James, D., 168
Salinger, G. L., 107, 110
Sample, H. H., 87–8
Sarma, G., 168
Sato, H., 98
Savvides, N., 68–9
Schiffman, C. A., 168
Schlömann, E., 23, 47, 57, 58
Schoek, G., 163
Schrieffer, J. R., 165
Schutz, R. J., 112
Schwartz, J. W., 77, 93, 100–1
Scott, B. W., 155, 161
Seeger, A., 78
Sergeev, O. A., 107
Seward, W. D., 92–3
Shanks, H. R., 173
Sheard, F. W., 27, 40–1, 89
Shmushkevitch, I., 172
Sidles, P. H., 9
Simon, F. E., 66

Simons, S., 60
Slack, G. A., 5, 48, 49, 50–3, 54, 55, 56, 90, 94, 99, 170–3, 174
Smirnov, I. A., 175
Smith, D., 112
Smith, S. C., 66, 164
von Smoluchowski, M., 64
Sondheimer, E. H., 141, 148
Sparks, L. L., 143, 152, 153, 158
Spencer, A. N., 40, 89
Sproull, R. L., 101
Sreedhar, A. K., 156
Srivastava, G. P., 66
Steele, M. C., 175
Steigmeier, E. F., 48, 173
Stein, S., 112
Stephens, R. B., 110, 111, 113
Stevens, K. W. H., 98
Stevenson, R. W. H., 40, 89
Steyert, W. A., 168
Stil'bans, L. S., 81, 174
Stoicheff, B. P., 110
Stuckes, A. D., 173
Sujak, B., 94
Sussman, J. A., 70
Suzuki, H., 79, 102
Suzuki, T., 79, 102

Tainsh, R. J., 150–1, 153, 157, 160, 162
Taylor, A., 101
Terry, C., 107
Tewordt, T. L., 165
Thacher, P. D., 58, 63, 64
Thellung, A., 1, 21, 70
Thomas, E. J., 168
Touloukian, Y. S., 143
Truell, R., 100
Tucker, J. E., 107
Turk, L. A., 77

Varma, C. M., 111
Vasil'ev, L. N., 170
Verma, G. S., 66
Vos, J. E., 60

Wagner, C. M., 93
Waldorf, D. L., 151
Walker, C. T., 58, 59, 77, 81, 90, 93, 99, 100–1
Walton, D., 99–100
Weinstock, H., 101
White, G. K., 5, 112, 147, 148–9, 150–1, 156, 160
Whitworth, R. W., 71

Wiedemann, G., 11, 116
Williams, D. J., 178
Williams, R. K., 54, 170
Wilson, A. H., 133
Woods, S. B., 147, 148–9, 163
Worlock, J. M., 99–100

Yaqub, M., 151
Ying, C. F., 100
Youssouff, M., 76

Zadworny, F., 94
Zaitlin, M. P., 101, 112, 113, 114
Zavaritskii, N. V., 8, 165, 166
Zeller, R. C., 105, 106, 110, 111, 112
Ziman, J. M., 1, 18, 21, 23, 24, 40–1, 53, 58, 66, 67, 73, 78, 79, 89, 109, 121, 124, 128, 133, 137, 138, 176–8
Zimmerman, J. E., 140, 159–61
Zimmerman, J. G., 8
Zych, D., 151

SUBJECT INDEX

absolute magnitudes of phonon conductivity
  alkali halides, 54–5
  dependence on atoms per unit cell, 50–3
  dependence on Grüneisen constant, 50–1
  dependence on mass ratio, 53–5, 173–4
  high temperatures, 50–6
  optic phonons, effect of, 54–5
  semiconductors, 173–4
accuracy of measurement, 9–10
adamantine crystals, 55–6
additive resistances, 38–9, 42–3
alkali halides, absolute magnitudes, 54–5
alloys, silicon–germanium, 91
amorphous solids
  crystallinity, effect of, 107–8
  specific heat anomaly, 110–1
  tunnelling states, 111–2
Ångström (temperature wave) method, 7–8
anisotropy associated with U-processes, 58–61
  helium, experiments on, 60–1
atomic ions in potassium chloride
  Rayleigh scattering, 90
  resonance scattering, 93–4

bipolar diffusion, 169, 172–3
  relation to 'ordinary' conductivity, 172–3
boundary scattering
  amorphous materials, 113–4
  Casimir theory, 61–2
  combined with internal scattering, 66
  first observed, 61
  high temperatures, 68–9
  Knudsen flow analogy, 62–3
  length effect, 65–6
  phonons in metals, 155
  polycrystalline materials, 63–5
  predicted, 61
  radiation analogy, 62
  specular reflection, 64–5
Brillouin scattering in glasses, 112
Brillouin zone for simple cubic lattice, 19

Callaway method, 36–7
  additive resistances, 38–9
  combined relaxation rate, 37
  Poiseuille flow not predicted, 39
comparative methods, 7
comparator, thermal, 9

copper, lattice conductivity in, 144–5, 153, 157
  deformed, 162
crystallinity, influence in polymers, 107–8
Debye conductivity theory, 13, 29
  infinite conductivity for continuum, 29
'Debye' method, applicability, 82
Debye specific heat theory, 22–3
diffusivity, thermal, 7–9
dislocations, phonon scattering by
  core, 77–8
  discrepancy with theory, 101, 163
  optical analogy, 78
  pinning, effect of, 102–3
  strain field, 78–9
  superconductors, 164
  vibrating, 79, 101–3, 164
dispersion curve, linear chain, 16–9
displaced Planck distribution, 33–5
Drude theory, 11–12, 116

electrical conductivity, expression for, 124
electrical heating methods, 6–7
electron–electron scattering
  experimental evidence, 150–1
  Lorenz number for, 151
  transition metals, 138
electronic conductivity
  in copper alloys, 153–4
  minimum, 135, 137, 145–7
electronic resistivity maximum, 135, 137, 145–7
electron–phonon interactions
  effectiveness, 131
  effective relaxation rates, 132
  electrons in impurity bands, 176–8
  energy and wave-number conditions, 128–9
  ideal electronic conductivity, 145–50
  maximum electron direction change, 130–1
  maximum electron energy change, 130–1
  Pippard theory, 139–40, 158–61
  scattering probability, 129–30
  U-processes depress conductivity 136–7, 145–7, 148
energy gap at Brillouin zone, 120–1
epoxy resins, loaded, 108
equilibrium (Planck) phonon distribution, 19

SUBJECT INDEX

exponential temperature variation of phonon conductivity, 56–9

F-centres, 81
Fermi–Dirac distribution, 119
Fermi energy, 119
Fermi surface
  in electric field, 123–4, 125–6
  fit into Brillouin zone, 121–2
  in temperature gradient, 125–7
Fermi velocity, 119
Fermi wave-number, 119
figure of merit (semiconductors), 169, 174
four-phonon processes, 50
free-electron gas, specific heat, 120

glasses
  boundary scattering, 113–4
  holes in, 114
  phonons in, 112–3
  radiation through, 105–7
  radio-wave analogy, 109
  ultrasonic attenuation in, 112
glass fibres, 113–4
Grüneisen constant
  conductivity, effect on, 50–1
  definition, 46
  harmonic solid, zero in, 47
  vitreous silica, 112

helium, solid
  anisotropy, 60–1
  isotopes, 87–9
  Poiseuille flow, 70–1
helium, liquid, Poiseuille flow, 71
high-temperature phonon 1/T law
  at constant volume, 49
  departures from, 48–50, 170, 173
  derivation, 46
  expression for, 47
  semiconductors, 170, 173
  volume effect, 48–9

ideal electrical resistivity, 133
ideal electronic resistivity, 133, 145–50
  at low temperatures, 147–50
interface resistance, 108
isotopes, 83–90
  additive resistances, 87–8
  enhanced scattering, 86–8
  in germanium, 89–90
  in helium, 87–9
  in lithium fluoride, 83–6
  in neon, 86–7
  Rayleigh scattering, 74–5

Klemens's cut-off procedure, 41

compared with variational procedure, 41–2
Knudsen gas-flow analogy, 20, 62–3

lattice conductivity in metals
  copper, 144–5, 157
  platinum, 144, 158
  silver, 156–7
  tin, 157–8
lattice resistivity in metals
  boundary scattering, 155
  dislocations in well-annealed metals, 155
  impurities, 155
linear chain, 16–8
lithium fluoride
  exponential temperature dependence, 59
  isotopes in, 83–6
loaded epoxy resins, 108
longitudinal heat flow methods, 3–5
Lorenz number
  Boltzmann statistics, 171
  gallium, 151
  ideal, 125
  platinum, 158
  temperature variation, 152–3

magnon conduction and scattering, 98–9
mass ratio of constituents, effect of, 54–5, 173–4
Matthiessen's rule, 128
  departures from, 141–2, 150, 154
molecular ions in potassium chloride, 91–4
  infrared absorption, 92–3
  resonance frequencies, 93

neon, isotopes in, 86–7
nomenclature, phonon and electron conductivities and resistivities, 133
non-metals with high conductivities, 55–6
normal modes for linear chain, 16–8
N-processes
  analogy with molecular collisions, 32
  ineffectiveness, 32–4
  in helium, comparison of rates, 88–9
  in lithium fluoride, comparison with ultrasonic attenuation, 85–6
  rate, population factor, 33–4

optic modes, 18, 54–5

paramagnetic crystals, 94–9
  coupled spin-phonon modes, 98
  model for scattering effect, 95–6
phonon
  flow, analogy with gases, 20
  focussing, 67–8
  frequencies, variation with temperature, 50

SUBJECT INDEX

group velocity, 20
wave packets, 20
phonons in glasses, 112–4
  as heat carriers, 113–4
Planck (equilibrium phonon) distribution, 19
platinum, lattice conductivity in, 144, 158
Poiseuille flow of gases and viscosity, 63
Poiseuille flow of phonons
  in bismuth, 72
  conditions for, 70
  gas flow, analogy, 69
  in liquid helium, 71
  in solid helium, 70–1
population factor in rate of change
  electron numbers, 129
  phonon numbers, 33–4, 45–6
precipitates, 99–101

radial flow methods, 5–7
radiation through glass, 105–7
Rayleigh scattering, 74–6
  transition to geometrical scattering, 77, 100–1
reciprocal lattice vector, 19
relaxation rates, thermal and electrical conductivity, 127, 132–3
relaxation time, 21

semiconductors
  absolute magnitudes, 173–4
  bipolar diffusion, 169, 172–3
  departures from 1/T law, 170, 173
  electrons in impurity bands, 176–8
  figure of merit, 169, 174
  pure, electronic conductivity, 171–3
silica, crystalline and vitreous compared, 104–6
silicon-germanium alloys, 91
silver, lattice conductivity in, 156–7
sodium fluoride, exponential temperature dependence, 58–9

steady-state methods
  comparative, 7
  effect of finite temperature drop, 4
  longitudinal flow, 3–5
  radial flow, 5–6
  self-heating, 6–7
superconductors
  diffusivity, 165–6
  dislocations, 164
  lattice conductivity, 165
  normal and superconducting states compared, 165–7
  thermal switch, 168

temperature wave (Ångström) method, 7–8
thermal comparator, 9
thermal diffusivity
  definition, 7
  determination, 7–9
  superconductors, 165–6
thermal switch, superconducting, 168
three-dimensional lattice, modes in, 19
tin, lattice conductivity in, 157–8
transitory heating methods, 8–9
tunnelling states in amorphous solids, 111–2

ultrasonic attenuation
  lithium fluoride, 85–6
  vitreous silica, 112
units of thermal conductivity, 3
U-processes, anisotropy associated with, 58–61

variational method
  additive resistances, 42–3
  entropy production, 24–6
  minimizing resistivity, 24

Wiedemann–Franz–Lorenz law, 11–3, 116–7, 125
  breakdown, 127, 132, 135

DISCARDED

COLLEGE OF THE SEQUOIAS
LIBRARY